KOSCIUSZKO

NICK BRODIE

BOOKS

Published in 2019 by Hardie Grant Books,
an imprint of Hardie Grant Publishing

Hardie Grant Books (Melbourne)
Building 1, 658 Church Street
Richmond, Victoria 3121

Hardie Grant Books (London)
5th & 6th Floor
52–54 Southwark Street
London SE1 1UN

hardiegrantbooks.com

 A catalogue record for this
book is available from the
National Library of Australia

Kosciuszko
ISBN 978 1 74379 401 2

10 9 8 7 6 5 4 3 2 1

Cover design by Nada Backovic
Typeset by Kirby Jones
Printed by McPherson's Printing Group, Maryborough, Victoria

 The paper this book is printed on is certified against the Forest
Stewardship Council® Standards. FSC® promotes environmentally
responsible, socially beneficial and economically viable
management of the world's forests.

KOSCIUSZKO

Dedicated to my friend
Jamie Ian Lyden
30 July 1981 – 17 March 2007
Requiescat in pace

Contents

Conventions

Kosciuszko is now spelled with a z, but for most of the period covered by this book it wasn't. Hence, while the modern form is adopted within the following pages, quoted material reflects earlier usage. As the Kosciusko Hotel burned down before the modern convention, its original spelling has been maintained.

Note that other modern usages also sometimes differ from quotations – e.g. Monaro ('Manaro') and Sakhalin ('Saghalien').

Preface

More than just our highest mountain, Kosciuszko reverberates in our national psyche in ways both perplexing and telling. Ours is an identity grounded in place and history wherein Kosciuszko has a special but easily overlooked part.

We seem to intuit that Kosciuszko's significance transcends height. It has an importance that is geographical and geological, but its effect upon us is also profoundly cultural. My ancestors lived up by Kosciuszko's side, and the mountain long called to me from that famous poem where a horseman rides out from Snowy River. I knew Kosciuszko before experiencing it, and I loved it without really knowing why. Kosciuszko was a part of me, and that was that.

But now Kosciuszko has called me back to address questions I had never thought to ask. Why does its very name ring in our souls? How has it shaped us?

Kosciuszko is at once inhospitable yet welcoming, untameable but familiar. It is a place of steadfastness amid the changing

1

days and rolling storms of history, yet history's changes stand out especially clearly upon its sides and summit. We need such perspective in this age of division. By taking us higher, Kosciuszko helps us look further. On Kosciuszko, we see our story play out. From Kosciuszko, we see our story reach out.

Representing a tactile truth that transcends us all, Kosciuszko silently speaks of our deep past. For uncountable generations the alpine world in which Kosciuszko is situated provided rich summer food and witnessed the unifying effects of ceremony. 'Bogong, Bogong', one Aboriginal group told a perplexed colonial enquirer, rubbing their stomachs for emphasis.[1] Following these people towards Kosciuszko he observed them gathering and consuming bogong moths, and caught a glimpse of Kosciuszko's ancient and unifying power. Drawing different peoples up into its heights, Kosciuszko brought many of Australia's First Peoples together. Before the Commonwealth, before the First Fleet, before the *Endeavour*, Kosciuszko was a place where nations met and stories combined.

For untold centuries, from every compass point, they gathered each Kosciuszko season. They travelled from the coasts and the great river grasslands, up through the hilly forests, into valleys being carved by icy snowmelt, towards the rocky and treeless peaks that marked the nearest points to the heavens. And there in country traversed by the ancestors since the Pleistocene, that glacial age of great floods, they found companionship and nourishment. Kosciuszko was, and still is, a place of great communion.

The mountain once had many names. Associations of specific words with exact sites weakened under colonial times, when whole languages were driven into silence, but some words remain to remind us that this apparent wilderness is a living human landscape. Australia's highest lake, nestled by the side of Mount Kosciuszko, is Lake Cootapatamba, a name bearing words both new and old. Other lost words may yet survive. One geologist in the 1850s was told by some Aboriginal people that Kosciuszko was 'called Piallow', a fact hiding in plain sight amid the newspapers of the past.[2]

But Aboriginal names were not mere labels. They typically spoke beyond location and feature to ideas and emotions. One account had the term 'Gibbo' used for these alpine regions, reportedly meaning 'snow mountains'. The same term also referred to the initiation of young men with white ochre, hinting towards the spiritual significance for which Kosciuszko stands.[3] Australia's Snowy Mountains were never just a place, they were an allusion to holy ideas.

In this we are not alone. Like Ben Nevis, Fuji or Kilimanjaro the mere whisper of Kosciuszko conjures a sense of country. Like Sinai it encompasses the mysterious workings of history and eternity, becoming a place in and of lore. To be Australian is in essence to have lived beneath the peak of Kosciuszko.

Wherever in the world we go, that word Kosciuszko brings us home. Perhaps more than any other Australian place name, Kosciuszko captures our distinctive accent. We have modified the consonants and dropped at least one vowel from the original Polish pronunciation. For generations we misspoke the word into something new – *kozzy-osko* – taking it out of its precise historical context and reclaiming it in the process.

This history does similarly, taking Kosciuszko out of historical tradition and narrative convention. Like a mountain summit its focus is narrow, but its sweep is broad.

*

We learn and cherish Mount Kosciuszko's facts, but they quickly guide us to feelings. It stands 2228 metres above sea level. It is the highest portion of Australia's Great Dividing Range, the stretch of mountains that effectively runs the length of Australia's eastern coast. A continent with a bit of a reputation for being flat turns out to have one of the longest mountain ranges in the world, the result of an intercontinental collision some 300-odd million years ago. But while many of us will struggle to recall our lowest point, our easternmost, westernmost, northernmost and southernmost points, our longest and shortest rivers, most schoolchildren in Australia can answer in a flash that Kosciuszko is our tallest mountain. That swift recognition comes from something beyond geography. It speaks of a history that is like a collective memory, buried but not quite forgotten.

In part this reflects the way that our mythic history played out upon Kosciuszko's sides. This is the land of Banjo Paterson's ballad *The Man from Snowy River.* The wiry little horseman who hailed from Kosciuszko steps from the shadows to ride brilliantly and vanish quickly. He stood in for an Australian everyman, a nameless icon of manly virtue and ephemeral humility. In his quiet confidence he challenged the bombast and hubris of our body politic, and remains a reminder that our greatest heroes are often nameless.

Yet while we feel its importance and our poetic heroes strut its sides, Kosciuszko gets left out of Australia's history as it is often told. It is almost as if one of our most iconic landmarks does not quite belong in the national story. James Cook never climbed the mountain, Arthur Phillip never settled upon it, and its earth never soaked the blood of Gallipoli. Above the narratives, beyond the platitudes, this most Australian of places often stays out of our story and remains but part of the scenery.

Professor Ernest Scott, for instance, after whom the Australian Historical Association's most prestigious award is named, gave but three brief mentions to Kosciuszko in his *A Short History of Australia*. It first featured as a point of comparison for the height of the Blue Mountains, and last as a geographical landmark to help readers triangulate the situation of the new capital of Canberra.[4] Between these merely comparative mentions was a passing note in a chapter about exploration. Scott pointed out in a sentence that Kosciuszko was named in 1840 by the Polish alpine explorer Paweł Edmund Strzelecki as he travelled to Gippsland.[5]

That Strzelecki was safely guided into and through the mountains by Aboriginal men like Charlie Tarra went unmentioned in the great professor's work. Yet without Tarra, the name Kosciuszko may well have stayed in the mountains. In that way, Kosciuszko speaks to the contradictions of our national story, and our failings of historical memory. It also shows that a small moment on the mountain, when read in clearer light, can reveal a greater historic vista and broader national truths.

Once we become alert to its cultural presence, the mountain points to its own ubiquity. Kosciuszko was, after all, demonstrably an important part of Scott's own world. In January 1939, for example, he was in Canberra giving a lecture to a meeting of the Australian and New Zealand Association for the Advancement of Science. A weekend excursion to Kosciuszko was a notable part of the proceedings.[6] The mountain provided space for scientific exploration, physical exercise, social relaxation and what would now be called professional networking. Kosciuszko was part of Scott's world, just not his vision for Australian history.

For Scott, like so many before and since, Australian history relied upon the activities of great explorers and important white men. 'This Short History of Australia begins with a blank space on the map,' he said in the opening line of his rather cartographically obsessed *Short History*, 'and ends with the record of a new name on the map, that of Anzac.'[7] His purpose was, he clearly stated, 'to elucidate the way in which the country was discovered, why and how it was settled, the development of civilized society within it'. It was basically chronological, and mostly interested in explorers and politicians and empire. And so for generations of Australian schoolchildren, if they were introduced to Kosciuszko's history it was most likely through reference to the first Europeans to see, name and summit it.

This history is different. It is not a typical then-to-now history of our nation's tallest mountain. It is instead a story upon the mountain, a journey into the world of Kosciuszko. We enter a moment that reveals the essence of a period and threads of history that traverse the world. We follow a trail from which we can look further backwards and outwards, seeing the mountain through time in a way that transcends sequence. We meet people whose attributes we recognise and whose relationship with Kosciuszko echoes in our own day.

This is a different sort of short history of Australia – one grounded by the heights of Kosciuszko. It is the story of a youthful and optimistic nation embracing their mountain, of a world of technological progress pushing the limits of human endurance, and of two men whose love of Kosciuszko transcended the barriers of class that would ordinarily divide them.

Our guides into this past were humble in the best possible ways. Neither were heroes in the traditional mould of explorer, governor or warrior. Instead, both serve as reminders of the depth that can hide behind a public face. After researching their lives in considerable detail, I became quite fond of them. They seemed like good blokes. In these times of political discord and historical amnesia we need to be reminded that our shared national story has many such people.

Evan Hayes was an ordinary Australian battler. Hardworking, likeable. Laurie Seaman was a worldly-wise American. Adventurous, affluent. Both were athletic and experienced cross-country skiers. In August 1928 they journeyed into that borderland called Kosciuszko where history and legend blur. Their trail leads us to an Australia that deserves to be better known. Between wars, this is a story of trouble in peace. Beyond politics, this is a story of unity in difference. And above easy platitudes of exploration and settlement, this is the story of a nation discovering itself.

Through them we meet Australians who show us their nation as it was: city and country, old and new, famous and forgotten. All were joined by historical circumstance, momentarily focused upon the icy wilderness by Kosciuszko's sides.

CHAPTER ONE

A Kosciuszko Childhood

Nobody lives on Kosciuszko. Each year some three million people visit Kosciuszko National Park, tens of thousands of whom climb the mountain before leaving. Even on the mildest of summer days the mountain weather can turn savage, so we visit but never stay. Eroding much like the great mountain itself, the stone cairns erected on its top keep crumbling, helped by the many feet of those who would stand upon Australia's highest point. Mountains are the proverbial icon of steadfastness, yet they are anything but.

When photographs first appeared of weird sculptures formed by wind and ice at Kosciuszko's summit, jagged spikes of frozen water blasted solid and sideways, the images seemed surreal. So unlike the popular image of a sunburnt country baked dry by the ages, Kosciuszko was set apart.

Yet many of our ancestors lived beside it, in the high-country plains below. These were the people of Kosciuszko, and Evan Hayes was one of them.

A Cooma Boy

Evan grew up in a space that has long captured our national imagination. Scenes of mountains fill our galleries because artists like Eugene von Guérard, Nicholas Chevalier and William Charles Piguent looked to Kosciuszko for romance and inspiration. In seeking Australia's own alps, they frequently met and depicted the people of the high country. In poetry such folk were immortalised, most especially through the rhymes of Banjo Paterson, and the Snowy Mountains were the inspiration and setting for some of our most popular novels, from Miles Franklin's *My Brilliant Career* to Elyne Mitchell's *The Silver Brumby*. Even for those of us who have never seen it, Kosciuszko has long been an important part of our collective cultural heritage.

But for many of us the connection runs deeper still. Up beside Kosciuszko, our culture intersects with kin. Many of us born in towns and cities across Australia can trace our families back into the mountains. The high-country clans that ran cattle near Kosciuszko were big, so I wasn't surprised when I discovered that Evan and my ancestral paths intersect.

Born many decades apart, near different rivers in different towns, we both hail from Kosciuszko. Between us, we triangulate a pull to the mountain country that seems to transcend both time and geography. Perhaps we carry the shadows of memories formed by our forebears: a certain thrill at the first sight of snow, a special delight in the stark freshness of thinner air.

Yet to get from shadowy feelings to hard historical facts requires a bit of luck. In fact, bad luck is often best, for tragedy is a great producer of documentation. One such instance of misfortune allows me to pinpoint our ancestral intersection with absurd precision. It took place on the road from Queanbeyan to Cooma, late in the nineteenth century. Evan's uncle Frederick Pooley owned the mail coach, and Bob Dean, one of my ancestors, drove it. When the coach crashed spectacularly in 1884, injuring Bob in the process, it also revealed a special link to the peak of Kosciuszko.[1] One of the passengers aboard was the surveyor Arthur Betts, after whom Betts Camp was named. It was the nearest hut to the mountain during most of Evan's lifetime, and one he knew well.

Although born in Paramatta in 1898, Evan was really a high country boy. Much of his family's history centred on Cooma, a gateway town to the Snowy Mountains region of New South Wales. His parents married in Cooma, and one of his siblings was born there in 1907. His aunt Emma Pooley ran Cooma's Royal Hotel. By the time Evan was a child the widowed Emma held the hotel's licence in her own right, running it until 1909, after which she moved to Sydney. About that time Evan's immediate family also left the district.

Like most childhoods of the past, Evan's was under-documented. But a sense of life in Cooma at the time can be gleaned from the letters of his contemporaries. Writing in August 1907, eleven-year-old Leila Nicholson corresponded with 'Uncle Jeff' of the *Albury Banner and Wodonga Express*'s Children's Page, explaining what her and Evan's home town was like:

> Cooma is situated about 264 miles from Sydney on the Southern line. It is a medium-sized town, and has hills all round, half of which are covered with trees and the other half are bare. The Murrumbidgee River flows within a distance of five miles from the town. There are two creeks at Cooma; they are called the Cooma Creek and the Royal Creek.[2]

Through Leila's regular correspondence with the Children's Page it is possible to follow the rhythms of the year and the changing seasons in Evan's Cooma. 'We are having nice warm sunny days,' she wrote in September, 'but we have dreadful dust storms as well'.[3] The willow trees were apparently blooming attractively.

The next winter she wrote of snowfall:

We had a lovely fall of snow here on the 23rd June – about 4 inches. It was great fun snowballing. The day after the fall was not so nice, for the snow was frozen and was very slippery.[4]

And in October she captured the vagaries of Cooma's climate:

We are not getting summer yet: it is a very changeable place. It comes real warm weather for a few days, and then it gets cold again; in fact it snowed about two weeks ago.[5]

Undoubtedly Evan had opportunities to enjoy the snow of Cooma too. In fact, he very likely got to experience a peculiar element of Australian alpine culture – skiing – a sport whose popularity grew greatly during his lifetime.

Although snow rarely settled long enough in Cooma to facilitate skiing, during the early twentieth century its residents were known to travel to the nearby village of Kiandra for that purpose. By then the sport had a long local tradition, dating from the gold rush years of the nineteenth century. 'Snow sports on a large scale were carried on at Kiandra 25 years before skis were used extensively in Switzerland', one local even claimed with some pride, pointing out almost as a sort of proof that some of the Chinese miners 'became very proficient'.[6] This was the sort of skiing that Evan knew as a child of the mountains. It was, in its Australian genesis, a community sport.

Beyond these ski trips, he probably travelled the wider area. On his father's side Evan had another high-country hotelier in the family. His grandfather managed the Rose Inn at Adaminaby, now especially known for its big trout. Another young correspondent from Cooma, albeit twice Evan's age, wrote of going fishing on the Snowy River in 1907 in company with his father. 'We camped in an old hut that night,' he explained, 'and of all the noises, I never heard the like. There were dingoes, foxes, wombats, bears [sic], owls, and numerous other animals and birds yelping, yelling, howling, barking, screaming, hooting, grunting and screeching in all directions'.[7] For country kids such camping adventures were never far away.

The Australian bush clearly left an impression. This same writer – identified simply as Albion, meaning 'Britain' – wrote lengthy accounts of playful bush adventures of a sort that Evan too may have enjoyed. Clambering up rocks to reach a waterfall was, for Albion, 'as if we were scaling a fort'.[8]

Through such references these letters reveal the intersection of experience, imagination and identity. Albion's very name spoke of an attachment to the Old World, just as the fortified rocks may have hinted at things medieval or fields foreign. While the children of Cooma inhabited a cultural setting that was distinctly Australian – Leila sent 'Uncle Jeff' a copy of Henry Lawson's 'Ballad of the Drover', for instance – their horizons were also broad.[9] She also reported reading the American anti-slavery novel *Uncle Tom's Cabin*, suggesting that everyone give it a go.[10] Yet for all the outside influences, these children could not escape the specific effects of the

Australian alpine landscape. Albion, for instance, wrote of walking to a cliff, finding the unforgettable silence of a frozen creek, and a waterfall turned to jagged ice.[11]

Nor were the children unaffected by the deeper Australian story of people within that landscape. Ironically, after watching a rain shower from a natural rock shelter, Albion reported that while heading home 'we crossed the creek again, got some reeds for spears, and said we were blackfellows'.[12] Among the waterways of Kosciuszko the Briton had become Aboriginal.

The Mountain Nearby

Mount Kosciuszko had a special place in the experiences of the children of Cooma. At once both proximate and distant, it was undoubtedly on their emotional horizons. Although likely too young to comprehend its full import, Evan lived in Cooma in a period when the tourist potential of Kosciuszko was being aggressively explored. In early 1906, for instance, when the Premier of New South Wales visited Cooma one local politician took the opportunity of holding a public meeting to advocate for 'the necessity for improving the road, and erecting an accommodation house at Kosciusko, which should be made a resort, not only for the people of New South Wales, but Australia.'[13]

Within a few years a 'Kosciusko Hotel' was a reality, although to an extent it grew from a joke. About a year after this call for accommodation the New South Wales Tourist Bureau opened what it called the 'Snowy River Summer Camp'.[14] This was 'a weatherboard building' designed to accommodate thirty people at a time. It was part of a concerted government effort to open the region to summer tourism. Seemingly tongue-in-cheek, the *Sydney*

Morning Herald referred to this as 'A Kosciusko hotel', and the name stuck.[15]

A slightly more advanced version of alpine accommodation was opened in time for the winter of 1909.[16] Like the Kosciuszko road, also constructed in that decade, this 'Kosciusko Hotel' proper was part of the government's effort to make the region accessible. Even the famed lake in front of the hotel, upon which holiday-makers could skate when it froze, was an artificial construction designed to develop the recreational potential of the area. While it was obviously geared towards tourists from further afield than Cooma, it seems unlikely that Evan's family would make no effort to travel at least this far to see the new hotel. After all, in summer it was a pleasant and easy drive from Cooma via Jindabyne. So busy was the road, in fact, that when the hotel was opened one of the local attendees quipped that,

> When I came here there were not enough of us in the district
> to have a chat on Sundays. Now it is nothing but Governors
> and State Ministers. When I came, I came in a bullock dray;
> now I am afraid to drive my buggy for fear of motor cars.[17]

It all seemed proof of the economic philosophy that if infrastructure were built people would come.

All this activity clearly made some mark on the local children. Young Leila also occasionally informed Uncle Jeff of prominent travellers who visited Kosciuszko:

> The State Governor, Sir Harry Rawson, is paying a visit to
> us this week, and will arrive by train next Tuesday morning.
> I think the school children are to meet him at the railway

station. I hope he gives us a holiday. His Excellency is going
for a trip to Mount Kosciusko and Dalgety. On his return he
will visit our school and the hospital.[18]

And he did. Cooma decked itself in flags, toasts to the King were
drunk, music filled the streets, and the schoolchildren – Evan
possibly included – sang for the governor and his party.[19]

Governors were important to this early history of Kosciuszko
tourism because their journeys helped promote the destination and
popularise the trip. Their celebrity counted for more than their
constitutional role, and readers from far-flung towns and cities
learned of Kosciuszko by following others in print. This had the
fortuitous effect of documenting the period when Kosciuszko was
also being aggressively explored as a winter destination.

The well-publicised 1913 expedition of Australia's governor-
general to Kosciuszko, for instance, tapped into a wider enthusiasm
for adventures beyond the comforts of the hotel and into wilder
reaches accessible only by dog sleds and cross-country skiing.[20]
After one day of skiing, the party reached the hut at Betts Camp,
roughly halfway to Mount Kosciuszko from the hotel, and 'a
pleasant evening was spent round the fire.'[21] But the mountain
defeated even the vice-regal expedition:

Next day old Kosciusko put on his worst behaviour,
and smote the hut with a howling blizzard of snow. The
following day the conditions had not improved. If anything
the snow fell faster ... With a furious gale behind it, the
party made the pace a cracker when it started back for
the hotel, and it arrived bronzed and bearded, like Arctic
explorers.[22]

As this account detailed, the dogs that helped pull loads of provisions to and from Betts Camp for cross-country skiers 'were recently presented to the Government by the Scott and Mawson expeditions'. Quite a publicity coup for the Tourist Bureau, these sled teams of Antarctic dogs attracted considerable attention in the 1910s, reportedly having 'taken very kindly to their new life'. The largest of the dogs, named Colonel, was even a local celebrity.

Beyond the Great Divide

Though Evan had demonstrable roots in the high country, and strong connections with Cooma, his family was also a transient one. Among the limited documentary evidence available, the birth of siblings in 1900 and 1909 place the family in Smithfield in suburban Sydney and then Lithgow, over the Blue Mountains west of Sydney.[23]

Their movements suggest an association with the railway line. Clearly Evan's father sought work where he could find it; and the iron tracks linked all the places of Evan's childhood. By 1911, when the family was still living at Lithgow, Evan's father, William Charles Hayes, worked as a miner. Up until that year their lives had been sparsely recorded. But then the documentary silence was shattered.

On the surface all seemed well in Lithgow, a country town with growing prospects. A plentiful supply of nearby coal fostered local industry. As the *Sydney Morning Herald* recorded in April 1911, it was 'The age of steel'.[24] Lithgow 'must inevitably become one of the greatest industrial centres in Australia', suggesting it could be 'the Pittsburg of the Commonwealth'. Only in January the Lithgow ironworks had tested some new equipment, and reportedly rolled the first railway lines manufactured in Australia.[25] It seemed like a year of momentous change. Once again the world Evan was growing up in was abounding in optimism.

While positive about Australia's prospects for becoming a manufacturing nation, free of dependence on imported goods and materials, the *Sydney Morning Herald* writer was concerned about the political impasse 'in regard to pay and hours of labour' that was an impediment to progress. Pointing to the contractual disruptions caused by recent strikes at Newcastle, the writer noted that the Lithgow ironworks was experiencing similar 'labour troubles'. The ironworks' previous owner had, they argued, 'struggled heroically to build up the industry, but his task was rendered impossible by the continual demands made upon him by the men.' With its characterisations of heroic businessmen and demanding workers, the paper was hardly impartial in its treatment of a complex topic. Evan's father may have felt differently about the circumstances.

Under relatively new owners at this time, the Lithgow ironworks looked to a prosperous future of making fencing wire, railway lines and various other steel goods. But the ironworks was also a focal point of tensions between labour and capital. Between factories, mines and quarries, the firm reportedly employed 'about 1700 men'.

Because Lithgow also hosted 'the Government small-arms factory', local production had a national security significance. As one commentator of 1911 noted, 'there is, of course, every need for us to accumulate a war reserve as soon as possible.'[26] Threats to production could thereby be seen as threats to national security. The unpacking and installation of some new American war-making machinery was part of the noteworthy action of the working week at Lithgow in late April 1911, as was the resumption of full-time work with the end of Easter holidays and the repair of a broken engine.[27] This all meant, therefore, that local labour politics had an acute international dimension.

Tensions were evidently simmering in Lithgow, but for the Hayes family life continued its normal rhythms. During the Easter break they likely attended church and socialised. Evan was a student at St Joseph's convent school, and his father was an active member of the Hibernian Society, a largely Irish-Catholic charitable and mutual aid organisation.[28] Their daily and weekly routines were of limited significance to the local press, and thereby remain mostly undocumented. But things were about to change.

Organised labour was making itself felt across the country in the decades either side of Federation. This was indicative of a wider network of agitation, where strikes could beget strikes, made more effective because of the inter-dependencies of industrial production firms and methods. In February 1911, for instance, a strike at the Carcoar mine started to threaten production at the Lithgow ironworks.[29] Workers at the blast furnaces reportedly asked for 'a shilling per day increase all around', coalminers for 'an increase of 4d [four pence] per ton in the hewing rate', and other changes to pay and working conditions. The employer responded to these requests by encouraging government prosecution against strikers, and publicly saying that 'I think the working people should treat a strike as they would a venomous snake – that is, to take every means in their power to kill it'.[30]

When non-union labour was brought in to the Carcoar mine the Blast Furnace Employees' Union at the Lithgow ironworks refused to unload any coal from Carcoar and went on strike.[31]

By August these tensions over pay and working conditions had reached crisis point. One Thursday a 'considerable commotion' disturbed Lithgow when a few hundred striking workers at the ironworks gate 'vigorously hooted the non-unionists' that had been brought in.[32] Similar events followed over subsequent days, and more non-union labour was gradually brought in to ensure production.[33] Eventually, as the *Lithgow Mercury* put it, 'the most serious disturbance ever witnessed in Lithgow took place at the blast furnace, and for two or three hours mob rule held sway'.[34] Demonstration had turned to riot.

As the non-union labourers changed shifts in the late afternoon of 29 August 1911, large crowds hooted, and boys soon started 'to pelt stones' at a group of non-union workers. The exact cause of the violence was unclear, and there were reports of mutual provocation. But regardless of the specific cause, once it got started the confrontation gained a momentum of its own. Months of frustration spilled over. The crowd rushed the ironworks.

Despite wielding batons, policemen were unable to restrain the crush:

> Over the railway sidings they ran, and swept in one big
> volume round to the east of the engine-room. Here some of
> the police took up a stand, but many of the crowd made a
> detour of the pig beds and came down round the furnace
> towards the engine-room.[35]

In this engine room some of the owner's sons and non-union labourers had barricaded themselves. The crowd commenced lobbing anything they could find at the building. Windows started breaking.

At one point the crowd sang 'Rule Britannia', evidently becoming most enthusiastic at the lines that 'Britons never, never, never shall be slaves'. And on the steps of the engine room, standing in front of this crowd, was William Hayes.

A Father's Troubles

From that moment Evan's life was irrevocably changed. Reports of events were contested, but a picture emerges amid the tensions. Evan's father can be seen getting caught up in great winds of history, offering hints of the difficulties faced by working families in an outwardly optimistic epoch.

According to one of the police sergeants, William addressed the crowd from the engine room steps.[36] 'Come along, boys, come along', William allegedly cried, 'we will have the ###! all out of here. They have got five minutes to come out, and two have now gone.' Ignoring the sergeant's attempts to talk him down, William then continued, 'Come along boys, they've only got another #####! minute.'

The sergeant then pulled his revolver, and ordered his men to do likewise. But in the crush it soon became apparent to the angry crowd that the police were hesitant to fire. 'Come along, boys', William supposedly yelled, 'come along, the ###! are not game to shoot.' With the police being hopelessly outnumbered, it was no surprise that they were hesitant to escalate the situation.

Essentially there was a stand-off at the engine room. The crowd wanted the furnace shut down for the night and the non-union labourers sent home. Those inside the building were, understandably, hesitant to trust themselves to the mercy of the crowd. As well as lobbing stones and bricks and iron at the windows, some had turned to other parts of the complex. The beds where some of the non-union workers slept had been destroyed, and the owner's new motor car was on fire.

Observing all of this were many townsfolk who had come to investigate the commotion. Among them was a reporter from the *Lithgow Mercury*. While the sergeant later testified that William 'was screeching like a maniac', the reporter's account was more nuanced. He could see that the union leaders like William were wedged between the crowd and the building, trying to balance the exploding currents of emotion with a bigger strategic picture:

> The unionists at the door continued to argue and to urge the non-unionists to come out, guaranteeing them safe conduct to their homes. 'Come off those steps,' cried the more impetuous among the crowd. 'Give them another chance,' was the reply of the unionists who were striving for a settlement.[37]

When it was finally obvious that no negotiation would succeed, another of the union leaders left the steps, and reportedly tried to calm the crowd. 'Remain passive and quiet, sit down quietly and yarn until the morning', he reportedly said, stating also that 'I am going to stand here in a sort of passive resistance until the morning.'

But in the early hours of that morning a trainload of police reinforcements arrived and the crowd was quickly dispersed. William was subsequently and peacefully arrested.[38]

In his defence William stated that 'I assisted Sergeant Burn that night, and I pulled a mounted policeman out of trouble.'[39] And while the police testimony placed William in maniacal rage at the forefront of the crowd, egging people on in a way suspiciously perfect for a charge of rioting, other witnesses saw William differently. 'Hayes was keeping the crowd back', one miner testified. 'Stones were flying pretty thick.'[40] Many other workers gave testimony in contradiction to that of the sergeant, especially about William's language. One reported seeing William with his hands up in front of the gun-wielding sergeant. Most witnesses said that William tried to restrain the crowd from violence.[41]

In his own defence at trial in the Bathurst Circuit Court, William admitted that he had been drinking before the riot and was 'very nearly drunk'. But he also testified that he had been drawn to the commotion rather than having been a cause of it.[42]

On at least two prior occasions William Hayes had gotten into trouble with the law, and both cases slightly foreshadowed this trial. He was fined in 1901 'for using obscene language at Smithfield'.[43] Earlier still, in 1899, he had been prosecuted for being 'drunk in Woodville-road' and 'resisting Constable Shields in the execution of his duty'.[44]

On Friday 13 October 1911 the jury retired for the night to consider whether William was guilty of rioting. They quickly determined in the affirmative, and the judge decided to make examples of William and two co-accused. All three were sentenced to 'be imprisoned and kept at hard labor in Goulburn Gaol for 15 months'.[45]

For a Father's Sins

William initially served his sentence at Bathurst. An early visitor from Lithgow commented that 'Hayes is engaged in the boot-making shop, and says he intends to learn the trade and open business in Main-street when he comes out'.[46] Part of the reason that people were taking an interest in him was because of continued agitation by labour organisations, which now focused on petitioning for the rioters to be released.[47] Many people saw the exemplary punishments as unfair, especially when two further co-accused had been found not guilty.

Politicking on his behalf eventually succeeded and William was released. A small report in the *Lithgow Mercury* described his return to Lithgow on 3 June 1912:

> Hayes was released this morning, and came to Eskbank
> with his wife and one child by this afternoon's passenger
> train. He was met at the station by a number of his old
> friends, and cordially welcomed. A coach was waiting, and
> he was driven to his home at Corneytown. Hayes, who looks
> well, says he was treated well, and received plenty of good,
> wholesome food.[48]

Undoubtedly William's trial and imprisonment were distressing for Evan and his siblings and mother, emotionally and financially. The effect upon the family must have been considerable, but all now seemed well. Yet the relief would not last.

One Wednesday in June 1912, about nine months after the trial, William and two other men were feted by a special dinner in Lithgow to celebrate their return.[49] There were cheers and toasts. 'The health of our three comrades', said one, while another concluded with wishes for 'health, long life, and our helping hand in the future'. When William took his turn to reply, he thanked the government, made a joke, and 'thanked the committee for the splendid manner in which they had cared for his wife and children during his incarceration'. The celebratory dinner went well into the night, and within a month William was back attending labour meetings. But tragedy was just around the corner.

Crossing some railway yards on his walk home from a miners' meeting, William stopped and waited for a goods train to pass.[50] The heavy thudding carriages drowned out the noise of another engine and William was struck down. One leg was severed completely. The other was badly mauled and almost detached too.

William's cries of pain and shock brought rapid help. 'What's up, Bill?' called a nearby miner as he ran to the scene. 'Oh,' William replied, 'my leg's off.'

Taken to hospital in shock, William died a few hours later. He was buried that Friday, with a large crowd in attendance at the funeral.[51]

Evan stepped in to the role of family provider.[52] A friend recalled that 'the responsibility of the whole family fell on his shoulders – a boy called on to do a man's share, and heroically he performed it'. And he did it by selling sound.

To Sydney

After the trial, imprisonment and death of William Hayes, Evan's family slipped from the pages of the newspapers. There were no new siblings born by railway lines to help track them, and the Great War chewed up much of the nation's print.

But by the early 1920s Evan re-emerged in the *Sydney Morning Herald*. He was now living in the city, in a period sometimes called the 'Jazz Age'. He had experienced the bush and then industry, and now he occupied the metropolitan world of culture, working for the Pianola Company Pty Ltd of George Street.[53]

Pianolas were popular self-playing pianos. Operators pumped foot pedals to make them play music from rolls of paper punched with holes. Compressed air triggered the keys, giving the impression that the pianola was being played. Owners could collect rolls of music, which helped to popularise the modern music market. In-store demonstrations were an important element of the sales, allowing shoppers to see the pianolas in action and hear the music. 'Let us play these over for you in our spacious demonstrating parlours', one advertisement suggested.[54] The diversity on offer is evident in some of the titles singled out for mention, like those for 1921:

'Sleepy Seas,' Hawaiian Waltz Song
'I Wonder Why,' Song
'You Ought to See my Baby,' Fox Trot Song'[55]

This was a period of classical concerts, old-time dances and jazz bands, so it is not surprising that there was a great variety of music on offer. This was especially highlighted in the annual Christmas sales advertisements, demonstrating the increasing role of music in consumer culture. In December 1925, for instance, the company claimed to have 'over 10,000 "PIANOLA" music rolls and song rolls' on offer.[56] 'Every class of music – classic, dance, popular, and ballad – is represented', it added, and at only a shilling each they were relatively inexpensive. Music tastes in the 1920s were varied, but the personal collection of music was a phenomenon that was here to stay.

This became even clearer during Evan's early years with the company. The Pianola Company rebranded itself as the Aeolian Company as smaller audio technology became more important, available and desirable. In the 1920s Evan's employer was pushing the Vocalion phonograph, which they claimed was 'the only phonograph that makes the singer sound in the same room with you'.[57] It reportedly reproduced anything from 'gay songs for the lighter hours' to 'jazz that sets the feet a-tingling!'[58] And its uses were not only musical, as the Vocalion allowed 'students to hear and study the pure accents of the true Parisian, the true Italian, the true

German', as the technology was adapted for language teaching.[59] In music and in words, Evan played a small part in bringing the wider world into Australian homes.

But it was not all smooth going. In 1920 the government fined Evan's employer £700 for breaching customs.[60] Evan also found himself in legal trouble when, in August 1921, he was sued for an alleged libel against Doris Henderson.[61] Nineteen-year-old Doris worked at the Government Savings Bank, and Evan had asked her to acquire two tickets to one of the bank's social dances. Evan received the tickets, but for reasons that remain unclear, he decided against going to the dance and returned them to Doris. She in turn sent them back to Evan, expressing some regret at having arranged the tickets in the first place, and telling him to whom they should be returned. In then returning the tickets to the correct issuer Evan allegedly made some comments about Doris in writing that were injurious to 'her character, and containing statements, which meant that she was a person not fit for men or women to meet at dances or to associate with socially' – or so the court report put it.

Ultimately Evan admitted to writing a letter, but denied it contained anything defamatory about Doris. The judge, perhaps wondering if this were not a case of youthful passions going awry, encouraged Evan to apologise to Doris. He did, and she reportedly accepted it. The judge then 'gave a verdict for the plaintiff [Doris] for one farthing, with costs on the lowest scale, adding that the verdict was not to be taken as a contemptuous one, but was given to enable the plaintiff to get her costs in the action'.

Fortunately, when Evan attended the 'Kentucky Dance' the following year, he did so without legal incident.[62] In fact his role as a salesman of musical equipment may have influenced his social circles, gaining him entry to parties and dances that the average country boy might not normally expect. In early 1923, the *Sunday Times'* social pages record him at a 'jolly morning tea party … Mr. Evan Hayes made a happy little speech, wishing the guests of honor bon voyage'.[63]

But such social mentions do little to reveal the complex reality of Evan's city life, especially in the heady 1920s. Evan had at least one more tragedy to endure, only documented in passing soon after his 'happy little speech'. He returned home from work one evening in February 1923 to find his mother unexpectedly dead.[64] Life was not all song and dance for the popular young pianola salesman, whatever impression the newspapers might have given.

Beyond the mere fact of socialising, the newspaper reports that place Evan at parties and functions do reveal some of the peculiar cultural aspects of his city. For instance, he attended the Wentworth Café's dinner dances in May and June 1923.[65] The latter of these was an American-themed event, and a large model Niagara Falls decorated the room. Evan also showed an interest in American culture when he danced at Sydney's Town Hall in October for the Olympic Ball dressed as the American film actor Harold Lloyd.[66] It is easy to see how he might have befriended the American-born Laurie Seaman when he also settled in Sydney within a few years.

In January 1924 Evan socialised at Vaucluse, he dined at The Ambassadors in Pitt Street in February, attended the Shamrock Ball for St Patrick's Day in March, and danced at a silver wedding anniversary in Glebe Point in December.[67] The following years brought more of the same sorts of activities. From dining at a function for the visiting New Zealand 'All Blacks' rugby team in 1925, to attending a musical performance in 1926 about 'cowboys and cowgirls' set on the border of Texas and Mexico, Evan's social world had quite an international flavour.[68] Mexican cowgirls, Irish balls and American costumes all show Australians drawing on the cultures of the globe for inspiration. While first raised in a world made famous by Banjo Paterson's *The Man from Snowy River*, Evan now lived in a setting perhaps more closely resembling that of F. Scott Fitzgerald's *The Great Gatsby*. He may even have seen the movie version of this American classic when it arrived in Sydney cinemas in 1927.[69]

Yet beyond the dance floors and dining rooms, Kosciuszko called to Evan. The city boy from the mountains was becoming a regular and well-regarded skier.

To Kosciuszko

One of those people who travelled through Cooma towards the Kosciusko Hotel and beyond, Evan was very active in the Sydney-based ski club scene of the 1920s.[70] At first he travelled to the mountains with the Kosciusko Alpine Club, and later with the Millions Ski Club. As early as 1920, Evan can definitively be placed at the Kosciusko Hotel.[71]

When the Sydney Millions Club made its first small expedition to the snow in 1922, Evan was already at the Kosciusko Hotel a week before, suggesting there was a period when he was active in two clubs.[72] And as well as being an active club member in at least two organisations, Evan also helped record club activities, and was one of the contributors to the first issue of *The Australian Ski Year Book*. This was intended as 'an annual record of the sport of ski-running in Australia', a small sign that the sport was maturing and gaining popularity.[73] In it Evan wrote a short account of the Millions

Ski Club, of which he was one of the founding members.[74] Part of the Millions Club, this Sydney-based organisation ran organised tours to the mountain ski fields each year, and at the time of writing his brief history Evan was its 'popular secretary'.

Evan had played a prominent role in forming the Millions Ski Club, and this allows his own story to come into even better resolution. 'Very few visitors to Kosciusko have not heard of Evan', one writer claimed with considerable justification, despite Evan's relative youth and working-class background.[75] Active in the competitive and adventurous aspects of the winter excursions, Evan won two skiing cups and came second in a third in the 1924 season.[76] Over subsequent winters he won or placed well in various competitions, revealing that his happy demeanour was matched with great athletic skill.[77]

As Evan put it in his short history of the club, these annual excursions were extremely popular:

> it is necessary for the club to reserve the whole of the
> accommodation of the hotel for the use of its members in
> what is now known as 'Millions Club Week.' On the second
> Friday of August in each year a special train is needed to
> convey the enthusiastic members, who, year after year, return
> to the snow-clad fields to take part in the winter sports.[78]

A photograph accompanying the short article shows a group of club members in front of the Cooma railway station. Evan is seated and smiling, front and centre.

Other photographs of Evan attest to his close association with the annual skiing holiday. In 1926 the athletic and bespectacled Evan was photographed beside a modified motor car, equipped with treads on its rear wheels and skis under its front wheels. 'This car is for use in taking passengers to the Hotel Kosciusko when the roads are under snow', explained the *Queensland Times*, before identifying 'Mr. Evan Hayes, a noted ski-runner' as the man standing beside it.[79] The inclusion of this photograph in a Queensland newspaper also demonstrated that Kosciuszko and the snow country were important even to those who lived far from it.

Another published photograph not only highlighted Evan's skiing reputation, but also provided a glimpse of his playful side. He achieved momentary fame in the Sydney *Evening News* after falling in the snow. The photograph showed him sprawled out, and clearly starting to get up. Someone managed a quick photograph of the 'expert on skis' during this moment of indignity. 'Novices, take heart!' said the caption. 'Even the best fall sometimes.'[80]

Evan no doubt took it all in very good humour. His recorded words are few, but those about him all suggest he was very amiable. Back in 1922, while staying at the Kosciusko Hotel, he dressed as a 'sheik' during a fancy dress ball held to honour the birthday of the King of Norway, highlighting his active participation in the social life of the winter season.[81] As well as being the club's secretary, Evan also reportedly wrote the Million Ski Club's song, titled 'Boorangalang'.[82] He had a 'cheery nature', said another of his friends, describing Evan as 'never one of a party, but rather the party himself' who 'conveyed to his friends nothing but joy and pleasure – just one wholesome treat'.[83]

This joyous demeanour was on particular display during one season, when Evan took a leading role in the reaction to a much-anticipated fall of snow:

One snowless year club-members played golf in the vicinity
of the hotel in the afternoon. That night welcome noises were
heard on the roof about 11 o'clock. It was snowing heavily.
A fancy dress ball was in progress, and many of the revellers,
including … Evan Hayes, rushed out to the Grand Slam and
came down on skis without waiting to change into Alpine kit.[84]

The image of Evan skiing with delight in his dancing gear surely
gives some impression of a man for whom pleasure in the snow
trumped any concerns with convention. In the best possible way the
man holidayed with a child's enthusiasm.

Yet Evan's characteristic gaiety potentially masked something
deeper and darker. 'The impression he gave was that he had no
cares and no worries', wrote a friend, 'but behind that easy-going
temperament lurked a life of sorrow.'[85]

In some ways, then, Evan's skiing holidays might have been more
than just a source of fun. It was a place where perhaps Evan sought
something deeper, drawing meaning from his past, and strength
within himself. He went beyond the gentle slopes of the hotel,
into the icy wastes of Kosciuszko, showing that the heroic age of
exploration had not passed. It was, after all, still only a generation
since Kosciuszko was first summitted by ski in winter.

Devoting considerable attention to the history of skiing in
Australia, that inaugural issue of *The Australian Ski Year Book* to
which Evan contributed includes an account of the first recorded
winter ascent of Kosciuszko on skis in 1897.[86] Such expeditions
echoed contemporary adventures in Antarctica and were followed
with similar enthusiasm. Writing for the *Sydney Morning Herald*,
Charles Kerry, the leader of that expedition, had described 'the

lonely little lake of Cootapatamba ... shown only by broken snow and drifts', and how at Kosciuszko's base there were large 'drifts 60ft and 80ft high ... chambered out underneath into immense caves'.[87] His account of the summit of Mount Kosciuszko was otherworldly:

> The surveyor's signal on the cairn was decorated with frozen streamers towards the west, and the whole of the summit was like a coral bed of brilliants in the glitter of the sun's rays.[88]

Just as the age of Antarctic exploration did not cease when the pole was reached, nor was Kosciuszko considered to have been conquered after 1897. The white, treeless landscape of Kosciuszko undoubtedly inspired growing interest among the adventurous, including Evan. In these early decades of the twentieth century more and more Australians sought their own experience of these Snowy Mountains. Ski clubs initially headed to Kiandra, but they were soon heading closer to Kosciuszko. As the sport gained more enthusiasts, more clubs formed, and some of the more capable club members were able to see this icy world. Among them was Evan, blending high-country skiing tradition with the urban sporting enthusiast. Thirty-one years after Kerry, Evan and his companions could still make new discoveries.

And he wanted to share his adventures for the benefit of his country. Shortly before making the trip to the mountains in 1928 Evan told a friend that 'I am going out amongst the big Alps stuff. I am not going to stay about the hotel but I want to encourage visitors to climb and enjoy the true Alps of Australia'.[89] Even nearly a hundred years later, that sense of adventure and generosity of spirit makes me like him. Despite Laurie Seaman coming from a very different background, and displaying a different demeanour, it probably made Laurie like him too.

CHAPTER TWO

American Liberty

It is a curiosity of colonial history that Australia's highest mountain is named after an American hero. Despite his noble birth, Tadeusz Kościuszko left Europe to fight for the cause of liberty. He travelled to America during its War of Independence and took up arms against Britain. A military engineer who had studied in Paris and thought deeply about the inherent equality of humanity, he later sought to buy the freedom of Thomas Jefferson's slaves. After winning fame and respect in America, Kościuszko also fought for the liberty of his own homeland, fighting Russia and Prussia for the freedom of Poland. Unlike so many of the colonial names bequeathed to the Australian landscape – usually after governors, politicians and their wives – Mount Kosciuszko stands for those universal aspirations and ideals embodied in Kościuszko. Inscribed in days of dispossession and convictism, the mountain pointed to something better, a different trajectory for human history.

In many ways, Laurie Seaman was a beneficiary of the world Kościuszko helped create. Born in 1894, a few years before Evan,

Laurie was a child of the city of Glen Cove. While the Sydney that Evan knew in the 1920s had a whiff of *The Great Gatsby* about it, it was the time and place of Laurie's youth that provided the inspiration for F. Scott Fitzgerald's famous novel. Laurie was not quite Gatsby, but he might have lived next door. Glen Cove sat on the north shore of New York State's Long Island.

Fittingly, Laurie's early years were comfortable. His father, William Seaman, was a businessman who eventually became mayor of Glen Cove. Named for his father, Laurie defined himself through his second name, so when his name appears in print it was typically recorded as 'W. Laurie Seaman' – echoing the convention famously adopted by the author Fitzgerald.

Curiously, Evan was also named William after his own father and, like Laurie, used his middle name. In another curious coincidence, Evan and Laurie were both names derived from their respective mothers' maiden names. But here the similarities start to disappear, laying bare the different circumstances of their early lives. Unlike Evan, 'W. Laurie Seaman' is very easy to trace.

Laurie studied at the Friends Academy in Locust Valley, New York, before attending Swarthmore College in Pennsylvania. There he earned a Bachelor of Arts, majoring in civil engineering. The college's annual publication, *The Halcyon*, reveals much about his activities during this time. In the second semester of 1915 Laurie was the senior class president and was a member of the Phi Kappa Psi fraternity.[1] He was involved in the college's sporting life, including basketball, football and tennis. At one time, Laurie was treasurer of the Delphic Literary Society, president of the college branch of the Young Men's Christian Association, and a member of a range of other groups including the Engineers' Club, the Joseph

Leidy Scientific Society, the Book and Key Senior Society and the Athletic Association.[2] He was socially active and well-connected.

After graduation, Laurie worked as an assistant superintendent construction engineer for the Turner Construction Company in New York City.[3] The company was owned by a Swarthmore College alumnus, which may have smoothed Laurie's path to the position. It was one of the most successful businesses in the country, made famous by its modern use of concrete. But Laurie was not there for long. When America joined the war in Europe, he resigned from his position and enlisted in the United States Army.

Under the caption 'Glen Cove boy is learning to fly at Texas school', a December 1917 edition of the *Brooklyn Daily Eagle* sported a photograph of a uniformed Laurie.[4] A brief summary explained that he 'has graduated from the ground school in the Aviation Corps, at the Massachusetts Institute of Technology', and that he was 'now at an army flying school in Texas, where he expects to win his commission'. Laurie became a flight trainer himself, ultimately serving as a second lieutenant.

With peacetime, Laurie returned to civilian life. He became a member of the Hempstead Harbor Yacht Club and was on its membership committee in 1919; he took a trip to Miami Keys with his parents in 1921.[5] He reportedly briefly worked for the Turner Construction Company again as a sales engineer, before taking on a role as 'topographical engineer and chief of field party' in his father's business in Glen Cove.[6] Certainly in 1920 Laurie was active as a prominent local engineer and veteran, being on the building committee for a local 'community clubhouse' being built in central Glen Cove with fundraising assistance from the American Legion.[7]

In 1921 he joined Alexander Graham Bell's telephone company as a commercial survey engineer and later the Highland Motor Company in Glen Cove 'as Manager-Treasurer during [its] period of incorporation'.[8] His early career is suggestive of the power of connection.

But Laurie's story also hints at an inner restlessness. Where Evan was constrained by family duty, Laurie had both opportunity and means to move about. Just as Tadeusz Kościuszko met the luminaries of his age, from Thomas Jefferson to Napoleon Bonaparte, so too did Laurie set out to meet the great people of his era. Just as Kościuszko's life helps map a world being wrought anew by fights for liberty, so too do Laurie's travels reveal the world again caught in an era of profound change and ideological contest. As Evan endured tragedy and then thrived in a changing Australia, Laurie became witness to the wider sweep of world history in a truly remarkable way during his own long journey to Mount Kosciuszko.

Pictures by W. Laurie Seaman

In 1922 Laurie commenced what a friend later described as 'a world-tour making economic studies for journalistic and private purposes'.[9] Travelling as a press photographer with the journalist Drew Pearson, an experienced traveller and writer, Laurie set out on a great adventure.[10] Through his photojournalism, Laurie attempted to understand and record the state of the world in the mid-1920s, capturing what in hindsight is clearly a crucial juncture of its history. This was a time when old empires were falling and new ones rising, when new ideologies like communism were threatening established social orders, and where the interlinking global economy was readjusting after the disruptions of war.

Leaving Glen Cove in June 1922, Laurie met up with Drew and headed towards Asia.[11] It was some months before their articles started to appear regularly in the American press, lagging well behind the travellers, but the text and photographs document Laurie's experiences with amazing clarity.[12]

. By August they were in Japan. Going straight into the boundary between empires, they set out for territory being militarily occupied by Japan. As Drew explained,

> Almost immediately after my arrival in Japan, I went to North
> Saghalien and Siberia which were at that time still in the
> clutches of the Japanese army. On the map it looks as if you
> could throw a stone from Japan to Saghalien Island. However,
> I left Tokyo on a sweltering August day, and after a week of
> steady travel found myself with summer clothes in a semi-
> arctic climate.[13]

Sakhalin Island had long been contested by China, Russia and Japan. At the time of Laurie and Drew's visit it was ethnically divided between people of Russian and Japanese backgrounds, largely because of nineteenth-century shared sovereignty arrangements between these two countries. Russia held the north, Japan the south. But in 1922 the whole island was under Japanese control. Laurie was seeing clear evidence of the rising military power of Japan in the years immediately after the First World War.

Surprisingly, however, the pair reported little trouble excepting one incident in south Sakhalin. Apparently, Drew explained, 'a bald-headed policeman dressed in a straw derby and a kimono wanted to arrest Seaman and me for taking pictures of a Japanese circus'. The Americans 'strenuously resisted arrest', but were unable to shake off the policeman or a gathering crowd, until 'as a strategic move' they split up:

> This discouraged him for he could not follow us both.
> However, by that time he had observed that our main purpose
> in life was buying chocolate, and he breathed easier. That
> was the one and only time that the Japanese were anything
> but exceedingly courteous to me during a long trip through
> occupied military zones where visitors are not usually
> welcome.[14]

It was a relatively rare mention of Laurie, generally a silent companion in Drew's stories. His activities are mostly charted by his own photographs. At some point on this island at the centre of world politics, for instance, Laurie raised his camera and photographed a group of dog sleds.

The import of their travels went beyond jocular encounters and mere illustrations of foreign places. 'The United States is interested in Saghalien only because of its reputed enormous reserves of oil', Drew quipped at one point, but even that comment had a deeper barb. He knew that Sakhalin was important because it 'is the only weak point in Japan's adherence to the Washington Conference'. This American-hosted conference between several countries with naval interests in the Pacific, including Britain, America and Japan, was largely designed to avoid an arms race and future war between these competing nations. In one sense it recognised Japan as a major power in the Pacific, with its own colonial interests, while also serving to limit Japanese expansion. Seeing Japan still occupying Sakhalin was a hint that they would not completely back down from colonial aspirations.

Fortunately the wider strategic signs were better in Siberia, which Laurie and Drew visited next. For seven weeks they 'travelled through almost every part of Siberia which was under the Japanese occupation'. In this tense frontier between empires Laurie photographed a Japanese soldier dressed for the Arctic conditions. He was not smiling and held a bayonet. But what the Americans generally saw was good news for the world. 'Everywhere the troops were leaving,' they noted, 'just as glad to get out as the Russians were to have them go.' They also observed that Japanese troops were leaving occupied China 'in accordance with the rules laid down at Washington'. While Sakhalin remained a notable exception, Japan appeared to be mostly holding to its agreement with the international community. Thanks to Laurie and Drew the world may have breathed a little easier.

Yet the pair did not stop at observation alone. Perplexed by 'the sudden right about-face in Japan's nationalistic policy', Drew arranged an interview with Premier Katō Takaaki when they returned to Japan. Katō had attended the Washington Naval Conference and had recently been elected prime minister. On 13 November 1922 he explained to his American visitors that Japan's military activity was dictated by historical circumstances, thinly alluding to the activities of European colonial powers in Asia.[15]

At some point during the meeting with Katō, Laurie took a photographic portrait. It was one of several such images of world leaders taken by Laurie during his time in Japan, subsequently syndicated throughout the newspapers of the world. This even included a photo of Crown Prince Hirohito, the future emperor.[16]

Yet Laurie was not just a photographic spectator. While Drew wrote the main articles, Laurie did pen some accounts of parts of his travels. One of these concerned an octogenarian named Itto Kojima, who could remember when the American naval officer Commodore Matthew Perry had arrived in a warship and demanded Japan open its ports to trade with the United States back in 1853.[17] Writing from Tokyo, Laurie recorded Kojima's testimony of this truly world-changing event:

> Itto Kojima was a boy of about fifteen, living in the hills,
> when Perry first sailed along the shores of Japan. Itto came
> down to the sea with multitudes of men, women, children
> and barking dogs, – all anxiously watching the ships as they
> hugged the shore in passing. Beacon fires on the mountain
> tops passed the word along to prepare for battle. Warriors
> burned their arrow points and sharpened various implements
> of warfare. Soon after Perry anchored his big black 'fire-boats'
> in Yedo Bay, the temples were stripped of their bells, which
> were set up in rows on the beach to resemble cannon.[18]

That a firsthand account of Japan's awakening to the modern world was recorded by an American mission interested in investigating Japanese disarmament practically drips with historical irony.

Nonetheless, the meeting with Kojima points to a genuine curiosity on Laurie's part. While in Japan he photographed various items of social and cultural interest, like slum areas occupied by low-caste Japanese, caged prostitute women, a tea factory, a locomotive factory and American-style billboard advertising.[19] These images helped articulate tensions within Japanese society between insular traditions and global trends, militarism and liberalism.

Amid the complex of issues facing Japan, one stood out.[20] 'The grim menace of Bolshevism which has spread its shadow from Russia around the world', Drew wrote, 'is making its sinister influence very decidedly felt in the Islands of Japan.' The mayor of Tokyo was reportedly worried because the people 'do not understand what they are reading but they are curious, and this is dangerous because they eat but do not digest'. This was a time when the possibility of worldwide workers' revolutions seemed a real prospect.

But the Americans were not opposed to communism on ideological grounds alone. There was one incident that had occurred in Siberia which Drew left out.

The Soviet Threat

Laurie wrote an account of his encounter with the communists of Siberia in a letter penned shortly after his release from Soviet custody.[21] Although treating his experiences with a certain literary flippancy, he also caught sight of the real dangers the world could face by an ideologically driven regime that preached equality while exhibiting brutality:

> HABAROVSK, Siberia, Nov. 4 – A husky Siberian guard
> blocked my retreat as I closed my kodak and turned to go.
> He wore the military uniform of the Far Eastern Republic
> and flourished a wicked looking rifle with bayonet attached.
> Neither of us could understand the other man's lingo, but
> I hastened to obey his gestures and entered the G.P.O.
> headquarters without further argument. I realized the prison
> of the G.P.O. or secret police located just across the street
> was evidently not considered a permissible target for hungry
> tourists going about shooting pictures everywhere.
> I had been spending several days in Habarovsk at the
> Junction of the Trans-Siberian Railway and the Amur River.

As visas were necessary before departing, I presented my
passport to the Government Party Order – G.P.O. – for the
customary acid test and 'Red' approval. This secret police
functioning under the Bolshevist Government was quartered
in a prominent three-story brick building just off the main
street. The entrance hall was dark and damp and I was barely
able to read a large red banner suspended on the side wall.
This fiery pennant bore the following inscriptions printed in
many languages:

'The Far Eastern Republic and the Russian Socialist
Federated Soviet Republic united. Workers of the world unite.
We will show the world a new way.'[22]

While waiting for his passport to be returned, Laurie had wandered
back outside, intending to photograph the building. But after pulling
his camera out and trying to frame a shot, he decided that the light
was not good enough so did not take a photograph.

On being escorted back into the building by the Siberian guard,
Laurie faced the prospect of real trouble:

A few words through a pigeon hole and I found myself inside
a small guard room. Frantic winding and loud talking through
an old style telephone (all Siberian phones require strenuous
treatment) – soon brought a tall, grizzly Russian officer
uniformed in black with an ugly revolver in his waist belt.
I placed my kodak in his outstretched hand and watched him
turn it about. He could not open it, yet haughtily said as he
laid it on the table:

'I'll keep this. It's a good one. We have no such machine
here, and I can make good use of it myself.'[23]

But Laurie was no pushover.

> Not thinking of any possible fate, yet begrudging the loss of a
> kodak purchased for this trip, I replied as calmly as possible in
> spite of his domineering manner:
>
> 'I have seen no warning signs and did not know that the
> taking of photographs was prohibited. I want my kodak, and
> shall stay here till I get it back again.'[24]

Laurie assumed that his translator might have softened this reply,
but it still prompted the officer to storm out of the room, ordering
Laurie to wait as he slammed the door behind him.

Several guards came and went while Laurie waited. They were
mainly occupied in getting or replacing keys kept by the telephone,
which gave the impression that the building had many locked
rooms. Only one incident stood out from this routine, at least for
the detained American:

> Soon a woman nearly in tears entered with some small
> bundles of food. It was for her husband who had been
> imprisoned without trial for five days. A guard pierced the
> brown bread many times with a long fork, tasted of the
> bottled milk, closely examined the surface of a dried fish and
> rolled the 'all day suckers' on the table. Apparently satisfied
> that no tools or weapons were concealed in the rations, he
> listed the items, rebundled and carried them away.[25]

Laurie was then taken further into the building and up some stairs
to another room. There he was questioned at length, and an officer
prepared a written statement for him to sign. Uncertain of its
accuracy, Laurie had to rely on his translator's affirmation that it was

a fair translation of his explanation about the visa and the aborted photograph. He was then taken back down to wait for another hour, 'thrilled ... [by] the realization that I was virtually under arrest'.

Eventually the commandant arrived. He was carrying Laurie's camera, and led Laurie to another room:

> Seating ourselves around a table in his office, the commandant bowed, produced a broad smile and said:
>
> 'First of all to show my good faith, let me give you an exit pass so that you may rest assured you are going to get out all right. And here is your kodak, Mr. Seaman.'
>
> Here the smile vanished as he continued: 'I am very sorry that this incident has caused you such an inconvenience. We are anxious to have foreigners visit us. We want them to study our people and our country. But when they probe too deeply into our private affairs, then we do not like it. Our guards are under strict instructions to halt any suspicious or unusual actions on the part of anyone. And you were lucky, Mr. Seaman, that you were not fired upon as you attempted to photograph our prison, for kodaks and Americans are quite a novelty in this country.'[26]

Trying his luck, Laurie thanked the commandant for his kindness, smiled and 'asked the commanding officer for permission to photograph his prison'. He wanted some pictures to prove his story. The commandant smiled and agreed. He even offered to arrange a pass and guide for Laurie.

When Laurie retired to his hotel for the night, he and Drew discussed what they had seen, heard and learned. They concluded 'that the Far Eastern Republic runs short of being a true democracy'.

With morning, and the prospect of better light, Laurie returned to the scene of his short imprisonment:

> Not trusting in the commandant's blanket permit and to prevent a second escapade with the guard, who might this time test the sharpness of his bayonet, I inquired at the office if my permission had been passed along the line. It had not and I was not surprised.
>
> I began to feel at home seated again in the guard room as I awaited the transmission of my request to his majesty two flights up. My attention was attracted to a pretty Russian girl who entered with food for a woman prisoner. I learned that her friend was a telegraph operator and had been placed in jail more than four days before on suspicion of sending messages to the 'Whites' in Vladivostok. This second instance that I witnessed in as many days thoroughly convinced me that the G.P.O. was a handy tool in the hands of the government for disposing of undesirables and such other individuals as caused them any inconvenience or embarrassment.
>
> The commandant's answer, short and snappy, soon came over the wire with fur flying in all directions: 'Under no circumstances will any photographs of the G.P.O. be taken either indoors or outside.'[27]

Soon thereafter Laurie and Drew left Habarovsk for the city of Harbin. They planned to visit China en route back to Japan.

Witnessing China's Humiliation

Laurie's encounter with China was short, and this passing contact is reflected in Drew's journalism. His observations were more general than incidental – broad commentary pieces rather than episodic stories of travel. But they still painted a grim picture of life in China during this period, as a formerly great empire seemed to risk collapse. If Japan was the rising power to Australia's north, China was one of Australia's more worryingly unstable neighbours. If it imploded, the catastrophe could destabilise the entire region.

Effectively divided, China was suffering from what outsiders could only see as political chaos. Interviewing key faction leaders in Shanghai, Drew and Laurie learned something of the complexities involved. Among their interviewees was the founding President of the Republic of China, Sun Yat-sen. Sipping tea while responding to queries about China's apparent disunity, Sun explained to his American guests:

> As a people we have been united for over two thousand
> years – longer than any other race in the world. Today China
> is passing through a spasm of readjustment. Every country has
> them. Your country had its civil war.[28]

He was seemingly highlighting that China's domestic security trouble was the problem of banditry, which Drew joked was a 'national pastime'.[29] In reality, it was a symptom of financial instability and governmental crisis on a huge scale. Yet while the Americans seem to have suffered no personal bandit-related difficulties – luckily, because kidnapping foreigners was lucrative – they did see aspects of the phenomenon. Laurie photographed a 'Captive Chinese bandit', locked in a small wooden cage on public display.

The American travellers were struck by the poverty they witnessed in China. Laurie photographed 'boat beggars', for instance, which Drew characterised as 'the poorest people in all

China'.[30] These were the people whose houses were their boats, huddled together on the rivers in great swarms of humanity. Yet beyond just poverty, there was great disparity. Laurie photographed some Chinese farmers scratching at the soil, an image that was later used to illustrate how Chinese farmers still lived by subsistence. The nation was easily threatened by starvation. Yet, as Drew pointed out, some Chinese people dined absurdly well. He noted how some sections of the population enjoyed shark fins – an expensive delicacy – while others had depended on shipments of American flour during a famine.

As the American flour indicates, there was an international dimension to this economic malaise. Illustrating China's poverty in contrast with its famed productive wealth, Laurie photographed Chinese children working in a cotton factory.[31] Many of the mills were English-owned, reportedly with workforces mainly composed of women and children. They worked on poor terms in dangerous and unhealthy conditions, reminiscent of the extremes of the Industrial Revolution. As well as having English operators, many such mills were now being run on the same model by Chinese owners. This was globalisation, evident well before the term was invented.

Laurie most likely visited this mill with Drew, who detailed their visit:

> Arising at three-thirty one morning I made my way through
> a cutting gale to the Yangtzepoo district of Shanghai. There
> I saw two great streams of women and children, half of them
> leaving the mills after a twelve-hour shift, the other half
> pouring into the mills for their daily twelve hours of service to
> the never-seen lord and masters.

I entered one mill, one of the largest cotton factories
in the Orient, operated by a British firm whose ships and
plants virtually control the arteries of Commercial China.
The foreman, a lanky Britisher from Manchester, ushered me
through the establishment.

In the spinning room were long rows of whirling spindles.
Before them stood children, from seven years up. The room
was warm and the children were stripped to their waists ...
Several of the women had babies strapped to their backs. The
babies of other women workers lay in baskets and boxes in the
aisles of the hot room.[32]

Christian Churches and missionaries were advocating to improve
such working conditions, but it was obvious to the Americans that
the main problem was an absence of state regulation. 'The workers',
Drew said, 'are entirely at the mercy of the employer.' Apparently
one mill had made annual profits exceeding a million dollars on
at least three occasions. 'This foreign industrial exploitation',
he argued, 'is the cause of intense bitterness on the part of those
Chinese Socialists and labour leaders who have heard of labor
conditions elsewhere in the world.'

Visiting the Land of the Long White Cloud

After leaving China, Laurie and Drew returned to Japan, then
headed to New Zealand.

The Americans attracted local attention for what the *Auckland
Star* called their 'tour through the Dominion'.[33] Responding to
queries about the purpose of his and Laurie's visit, Drew told the
local reporter:

You have no idea how many people are curious to know more about New Zealand. Several editors of well known magazines and newspapers told me before I left, 'We want to know the truth about New Zealand. Is it the white man's Utopia? How efficient are its railroads, and how satisfactory is the Arbitration Court?'

The railways of New Zealand were state-owned and run, perplexing Americans familiar with a network of private railway companies. The Court of Arbitration, dating from 1894, could hear industrial disputes, set wages and determine standards of employment.

With typical enthusiasm, the Americans headed to Wellington and sought an interview with the prime minister, William 'Bulldog' Massey.[34] Laurie got to photograph the prime minister, and the interview process proved as much of an education as the interview itself.

'It usually takes anywhere from a few days to a month to interview a prime minister', Drew noted. 'You have to submit a list of questions in advance and study and dress for the occasion as if you were going to be married.' But much to their surprise, on 6 February 1923, the day before the opening of parliament for the year, they were ushered in to meet Massey after only about fifteen minutes waiting. Fumbling a little before the 'dominant, quick, eloquent' prime minister, Drew asked several questions about international relations and trade. His final question elicited a typically swift rebuke of American policy:

'What,' I asked, 'can we do to strengthen American–New Zealand friendship?' The answer came like the crack of a whip:

'Take off your tariff.'

'Take off your tariff,' he repeated, 'and let us deal with you. We want to sell to you. We want to buy from you. We can't do either. We did ship you a little butter last Christmas, but not enough to butter your bread for one day. Take off your tariff, and we can trade with you.'[35]

Here the interview ended.

Beyond economics, the Americans were deeply interested in wider cultural familiarities. Drew quoted a semi-serious article about American women from a February issue of the *New Zealand Herald*, which played on stereotypes about Americans. 'The American man feels that he is proving himself a great fellow by sticking obstinately to his blunder, once he has made it', the article asserted in a section Drew did not quote.[36] He did quote in part this good-humoured dig at American men, which played on their reputation for brashness:

Argument is entirely wasted on an American man, because he always starts out with the assumption that there isn't anything he doesn't know. This makes conversation with him somehow difficult. He talks voluminously and at great speed, and when he gets out of breath he waits for your applause.[37]

Drew ended the quotation before the applause part, perhaps aware of the sensitivities of his American readers.

Taking such obvious pleasure in the sport of international rivalry was itself evidence of a sense of cultural connection, if not homogeneity. While the people of the two nations were obviously different, the similarities were overwhelming for the Americans:

To the New Zealander the United States is the nearest white man's country; his mail and his cablegrams cross it coming from England; he or some one of his friends has been there. He reads all about us in his own newspapers; he buys our books and magazines; he sees us in the movies. Put an American down suddenly in a small New Zealand town and he senses the atmosphere immediately. He has to rub his eyes to make sure he is not in an American main street. The frame houses are built in the same hurried, inartistic style; the streets are just as straight; the cracker-barrel bums stand on the corner in their short sleeves and chew tobacco just as they do in Arkansas. The farmers park their dusty Fords in the middle of the wide street while they buy a mowing machine made in Racine, Wisconsin.[38]

For Drew, New Zealand deserved its reputation as a 'white man's Utopia'.

But Laurie turned his lens to a part of the New Zealand story left out of Drew's articles. He photographed Māori people, including a woman riding a bicycle with a child tied to her back.[39] While the *Pittsburgh Daily Post*'s caption flippantly noted she didn't need a train, the photograph itself captured the woman's dignity. There is even a hint of mutual human curiosity, evidenced by the way the child was looking directly at Laurie as he exposed the film. The 'white man's Utopia' was not so white after all.

The *Pittsburgh Daily Post*'s image and caption of the bicycle-riding Māori woman was meant to contrast with another of Laurie's photographs: a steam engine belching thick smoke.[40] This machine illustrated Drew's assessment of the public rail system, which he found irritatingly slow. He seemed even more annoyed by the laconic attitude of the staff and passengers when they faced such delays with a patience he clearly lacked.

While Laurie continued to be Drew's largely undocumented travel companion, Drew himself can be followed in detail because he spent a considerable amount of time journeying around New Zealand by train, giving lectures about Asia.[41] Laurie's Kodak undoubtedly captured the snapshots that illustrated Drew's lectures, although to what extent Laurie travelled within New Zealand with Drew is unclear.

Crossing the Tasman Sea

By April Drew was in Victoria, giving his 'Snapshots of the North Pacific' lecture in Sale on 25 April 1923.[42] In May he delivered it in Geelong, before taking the show to Hobart and Launceston in Tasmania.[43] Laurie's photographs evidently formed part of Drew's show.[44] Their arrival in Australia is particularly interesting for they were serving two roles: bringing the world to Australia, while also capturing Australia for the world. Curiously, their observations soon intersected with those historical forces that had impacted on Evan's early life.

Drew quickly observed what he characterised as 'a crisis, in which big business and the workingman are engaged in a bitter class struggle'.[45] A decade after Evan's father had been imprisoned, Australia was still struggling with industrial relations. Fresh from witnessing the treatment of workers in Japan, China and New Zealand, Drew was shocked by what he saw in Australia. He was horrified by the distribution of land ownership:

much of Australia's best farmland is held by wealthy sheepmen
and cattlemen in holdings which sometimes run up into millions
of acres. In the meantime, armies of unemployed, unable to
seek a living from the soil, flock to the cities and glut the labor
market. In Melbourne ... I saw hundreds of men sleeping in the
parks and under bridges, wrapped in newspapers.[46]

Surveying the political malaise caused by ideologically driven
governments, both left and right, Drew pointed to the phenomenal
rate of industrial dispute in the Australian coal industry, and
detailed the strategic advantages that large landholding farmers
gained through monopolising water access. Photographs of Broken
Hill and a camel caravan 'Hauling wool across Australian desert
for shipment to U.S.A.' illustrated his article, suggesting that Laurie
travelled into western New South Wales. While Laurie had perhaps
looked for a romantic touch in this last image, Drew's general
conclusions were stark:

Australia's labor problem, therefore, is inextricably tied up
with ownership of land and quantity of rainfall. But both
labor and capital remain blind to these root factors, and while
they continue talking about a Red Objective on one hand
and private enterprise on the other, the gap between them
continues to grow.[47]

He was much more optimistic about the labour situation he
encountered in New Zealand.

Yet while aware of its internal political sepsis, Drew was
conscious that Australia played a much more significant role
within the global community than New Zealand. In advance of the

Imperial Conference in Britain in late 1923 he secured an interview with Australian Prime Minister Stanley Bruce.[48] It was most likely conducted in Melbourne, sometime in August, and Drew seemed impressed by the thirty-nine-year-old leader. Bruce contrasted sharply with the former prime minister, William Hughes, who, Drew pointed out, used an 'ear trumpet' as an aid to hearing – and as a theatrical device for avoiding questions he did not like.

Bruce told Drew that the upcoming conference would be 'a turning-point in the development of the empire'. He was going with the clear objective of getting Australia to have a say in determining imperial policies:

> 'It is exceedingly embarrassing,' he went on very earnestly, 'for Australia to be called upon to send her ships and her soldiers and her money to fight a war on the other side of the world in which she has no concern and in the starting of which she has no part. Australia must know where she stands on this matter of Empire foreign policy. The final outcome of this one point will either draw the dominions closer to the mother country than ever before, or separate them more widely.'[49]

These were strong words coming from the conservative side of Australian politics. After the First World War, Australia wanted recognition of its maturity as a nation. 'I believe that this conference will recognize that each dominion is now an equal member in the British family of nations, with a voice equal to Britain's in the deciding of empire politics' the Australian prime minister told the American journalist. When asked about Australian relations with America, Bruce answered that they 'have always been the most friendly, and now I know that the somewhat similar characteristics, and similar ideals shared by the two countries will bring us even closer in the future'.

By now, however, Laurie had left Australia. Although credited with the photograph of Bruce, he must have taken it many months earlier. While Drew stayed on in Australia, Laurie continued around the world. He had not yet seen Kosciuszko, but as he steamed out of Sydney, the reason he later would was on the same ship.

Miss Chrissie Bell

The person who most guided Laurie towards Kosciuszko was not Evan Hayes, but Christiana De Rossi Bell. Generally known as Chrissie, and sometimes as 'Topsy', she, and Laurie too, was on the London-bound steamer *Ormonde* as it departed from Sydney's Circular Quay on 10 March 1923.[50] Chrissie commenced her voyage with at least one friend on board.[51] This was Joyce Russell, who was travelling first class with her well-to-do, well-connected parents. Chrissie travelled otherwise unaccompanied in second class, heading off for an English holiday with relatives.

Chrissie and Joyce knew each through the Voluntary Aid Detachment. Active during and after the First World War, the VAD was a woman's auxiliary organisation, and Chrissie's association helps chart something of her background.

A few weeks before departing Sydney, for instance, Chrissie received 'parting gifts' at a tea party held in her and Joyce's honour.[52] 'The tables were effectively decorated, and a merry time was spent', noted one report of the event. But the merry time reflected a more serious context.

The tea party was held at the Blinded Soldiers' Cafe in Pitt Street in Sydney, part of the charitable legacy of the war. In an interview with the cafe's 'commandant', conducted about the time that Chrissie was feted there, it was reported to be a popular place for drinks and meals and well supported by the wider community.[53] Some of the patrons even brought gifts of flowers for decoration or eggs to help reduce the running costs. 'For instance', the commandant told the reporter, 'there is one dear old lady who often brings us in a basket of apples – that means that apple pie figures on the menu that day, at a trifling cost to us.' She also mentioned another regular:

> One of our most consistent friends has been the 'Parsley Girl.'
> She is a young girl, who goes past this shop every day on her
> way to her office, and regularly every morning she brings in a
> bunch of parsley, grown by herself, as her little offering to the
> blinded men.[54]

Such aid was greatly welcomed as it helped increase the profits, 'every farthing of which goes to the blinded soldiers'.

By the early 1920s Chrissie was active in this charitable scene, a connection that grew out of her own wartime service. She was sufficiently well known in this regard within Sydney to attract a personal departure notice in early March:

> Miss Bell, only daughter of Mr. and Mrs. E. W. Bell, of
> Burwood, will leave by the Ormonde on a trip to England.
> Miss Bell, who was a Voluntary Aid during the war,
> has continued her work in the Lane Cove and Record
> Detachments, and has given valuable help to Red Cross
> convalescent homes and the Blinded Soldiers' tea shop. Miss
> Bell has been entertained at many farewell parties by her
> friends and co-workers.[55]

Chrissie was evidently a compassionate young woman, but her wartime service also stemmed from family tradition. One of Chrissie's grandfathers was Colonel Albert Bell, who had fought in Italy with Garibaldi. She also had a grand-uncle, another colonel, who 'in addition to serving with Garibaldi, fought in the American Civil War under [G]enerals Miles and Grant, and afterwards raised, at his own expense, a regiment of yeomanry, which served with distinction in South America'.[56] Furthermore, her mother was reportedly descended from 'one of Nelson's captains, who commanded the HMS *Fourdroyant* [sic] at the Nile and Trafalgar'.[57]

Unsurprisingly, given this family tradition, Chrissie also had a brother in the war. Roy Bell enlisted in 1915 and served as a gunner in the field artillery in France.[58] Like many young men he diarised his service and experiences, recording his voyage to Egypt and his many visits to churches, pyramids, the Sphinx and so on.[59] Once in France he was regularly on active artillery duty, shelling the enemy and being shelled in turn. He was frank about some of the devastation. But on 22 September 1916 Roy recorded an unusual incident where 'an intimate friendship sprang up between otherwise deadly enemies':

> At one part of our front [the] line is only about 15 yards
> from that of the Germans and so you have to speak in a
> whisper. I saw one curious thing which was that one of our
> periscopes had been put exactly opposite to one in Fritz['s]
> trench, and as I looked in on several occasions I could see the
> Germans looking at me, so that we had a good look at each
> other. During the morning he threw over some cigarettes and
> the boys in return showed him a tin of bully beef ... and he
> nodded, so over went two tins.[60]

It was the kind of thing that might have given readers hopes for peace. But it wasn't to be: Roy was killed in action a few weeks before Christmas 1916.

A few months earlier, in June 1916, Roy had made a passing mention of having received 'four parcels from home per the Comforts Fund'. Such funds were often organised by women, especially relatives, and Chrissie was likely to have been active in this endeavour. In early 1917 one 'Miss C. Bell' was elected assistant secretary to the 2nd DAC and 22nd Howitzer Brigade Comforts Fund.[61] This organisation had a shop in George Street, Sydney and ran stalls at opportune events like the Agricultural Show. The fund committee also ran their own fundraising events, like one held in April 1917 in which singer Kate Rooney performed her last Australian show before moving to the United States 'for home and hubby'.[62] Lectures on Papua, the 'Life in the Never Never', and a display of Japanese martial arts also featured in the program. In 1918, 'Miss C. Bell' was 'presented with an entrée dish, and received the good wishes of the committee for her forthcoming marriage'[63] – but since Chrissie was still single in the years that followed, this may have been someone other than her. Or it might have been a newspaper misidentification, or a cancelled wedding.

As with Evan, Chrissie's earlier years are harder to get at. It was likely, for instance, that she was the 'Miss C. Bell' who dressed as 'Folly' for a dance in Balmoral in 1912,[64] and the 'Miss C. Bell (Sydney)' who helped staff a 'Fancy and Work Stall' at the 1913 Jones' Island Flower Show in northern New South Wales.[65] (Her father often travelled to Lismore for work.) But from 1919, she was more easily identified. This shift occurred with the report of an accident. The *Daily Telegraph* briefly summarised the incident:

> While Miss Bell, daughter of Mr. E. W. Bell, of Burwood,
> was driving her mother and her aunt, Mrs. H. Bradley, in a
> motor-car on Thursday afternoon, they came into collision
> with a much heavier car at a street crossing. The occupants of
> the smaller car were all thrown to the roadway. The two elder
> ladies were conveyed to the local hospital in an unconscious
> condition. Miss Bell, although suffering from a broken
> wrist, remained by her mother's side at the hospital all night.
> Favorable progress is being made by the patients.[66]

While the crash revealed Chrissie dutifully and even stoically attending her mother, it also puts her behind the wheel of a motor car in Sydney in 1919. To an extent it marks the sort of social and cultural progress that the war had facilitated, although it was also a function of her growing into young womanhood. Her driving would not have been particularly remarkable but for the crash.

As well as accidents and charitable activities, Chrissie was apparently musically inclined, and the threads of her youth gradually form a picture of her enjoying the early 1920s. She was very likely the 'Topsy Bell' who contributed some musical talent towards a 'concert and euchre party' held at the Sydney Voluntary Aid Detachment Club in 1922.[67] Attending a big farewell party for naval wives prior to the *Ormonde*'s departure, Chrissie was also certainly the Topsy Bell who was part of the 'musical programme' arranged by the Voluntary Aid commandant who had recently hosted the tea party in her honour.[68] In this capacity, she may well have even known the pianola salesman Evan Hayes. But while surviving newspaper reports can map out some of her activities, they do not place the two together, at least not in any regular fashion. Chrissie and Laurie, on the other hand, can most decidedly be placed together.

To England, via the Indian Ocean

The *Ormonde* travelled via the southern capital ports of Hobart, Melbourne, Adelaide and Fremantle.[69] Each departure was a festive affair, typified by the send-off the *Ormonde* got when departing Melbourne on 17 March 1923.[70] 'There is nothing of the dreary weariness of saying good-bye at a railway train', the *Argus*'s reporter claimed. Instead, the scene was filled with 'buzz of conversation and excitement' and an 'atmosphere of interest and goodwill'. It was tourist season, and Australians with the means to do so were flocking to London.

The cruise was the first leg of the holiday, and by the afternoon of 23 March the ship was departing Fremantle, heading into the Indian Ocean.[71] Reports from similar passenger ships indicate that the Indian Ocean voyage was often filled with fancy dress balls and various social events and entertainments.[72] By 1 April the *Ormonde* was at Colombo in Sri Lanka (which the British called Ceylon), and ten days later at Suez in Egypt.[73] Brief shipping reports do little to record the nature of the voyage, but later accounts hold that 'a romance … began on shipboard' between Laurie and Chrissie.[74] They likely eyed each other at dinner functions, talked as they strolled about the ship, and danced during the evenings.

But the romance had one hiccup, brought about by Laurie's journalistic duties. Chrissie and Laurie's shipboard romance was interrupted when Laurie had to photograph Mahatma Gandhi.

Because the *Ormonde* did not stop at India, Laurie probably disembarked in Colombo. Remaining aboard, Chrissie managed to avoid catching measles during the epidemic that afflicted the ship. Her dashing new friend was clearly not far from her thoughts as the ship passed Suez in early April.[75] While the ship was at Egypt, she took a guided tour to the Sphinx.[76] Her guide reportedly predicted that she would marry Laurie – which more than suggests that Laurie's name was being mentioned.

After Egypt the *Ormonde* quickly crossed the Mediterranean, visiting Naples, Toulon and Gibraltar, before reaching Plymouth and then London in late April.[77] Chrissie was recorded among the disembarking passengers at Plymouth.[78]

While in India, Laurie managed to take his photograph of Gandhi, which was published a few months later.[79] He arrived in England a few weeks later by the *Narkunda*, another liner from Australia, which had stopped at Bombay en route.[80] That window had given Laurie time to complete his photographic odyssey of Asia before again meeting up with Chrissie.

An Atlantic Connection

As Chrissie and Laurie each stepped onto British shores their story might have remained obscure, but for the timely arrival of Laurie's parents. Prior to leaving America back in 1922, Laurie had arranged to meet them in London. Their travels help document his own, and Chrissie's too. The senior Seamans left New York in late June 1923, and after meeting Laurie they visited friends and relatives in England. A report of their activities was published in the *Brooklyn Daily Eagle* back in America:

> they saw something of the English country life, they toured
> the north of England, visiting the English lakes. They also

enjoyed the Trosachs in Scotland, the Scottish lakes, Oban
in the Highlands of Scotland, and the birthplace of Mrs.
Seaman's grandfather, as well as the famous resort Bettws-y-
Gped, which is so attractive to artists.[81]

The Seamans had also met Chrissie, who was now travelling with
them. 'In England the party visited Miss Bell's relatives,' the report
also mentioned, indicating that 'she will accompany [the Seamans]
to New York, returning to her home in Sydney by way of the Pacific
Ocean.' By October an engagement had been announced.[82]

Before that, however, there was Europe. Laurie reportedly
'persuaded Mr. and Mrs. Seaman to fly with him to Paris from
London'.[83] It seems likely this was their first experience of flying,
which was given much more attention in a letter Laurie's father
wrote about the experience than the destination itself. He was
impressed at how 'smooth' the flight was, and that none of the
passengers got 'seasick'. The party checked into the Grand Hotel in
Paris on 8 August.

Having reached the city whose ferment of ideas had so affected
Tadeusz Kościuszko roughly a century and a half earlier, Laurie
too may have been affected by the cross-currents of his own age,
brought into great relief by his own adventures. His European
experiences over the coming weeks remain relatively slimly
documented, however, although some final photographs suggest he
still travelled widely. Illustrating an article about Czechoslovakia for
the *Pittsburgh Daily Post*, Laurie provided images of gothic bridges,
rustic peasants and 'picturesque mountains'.[84] These mementos
offer a tantalising glimpse of how, having crossed the Pacific and
Indian Oceans, he also traversed some of the mountains of Europe.

When, like Kościuszko, he sailed from Europe for America, Laurie did so with an Australian lover on board. He had travelled more than most men alive and seen the tides of history ebbing and flowing into the earth's far-flung corners. Yet only now was his great migration beginning. He may not have known it, but Mount Kosciuszko was calling to him.

CHAPTER THREE

The Millions Club

By the 1920s Mount Kosciuszko was part of the cultural lexicon of Australia. It was whispered in poetry and blandished in advertisements, and against it were measured the breadth and length and height of the only nation on earth that also occupied a continent. Without ever visiting it, no child of Australia could escape Kosciuszko. Nor could any migrant to Australia hope to understand their new home without at least sensing its cultural significance.

In large part this derived from geography and topography. Australia's highest mountain served as a regular byword for snow, as it did in Brisbane in 1925 when *Truth* railed against 'a drunken orgy' in Victoria Park one night.[1] Apparently some of the language sung during this episode 'would have sizzled the top of Kosciusko'.[2]

Less scandalously, a chilled white dessert took its name from the mountain. 'Kosciusko Fluff' was essentially an unbaked pavlova, appearing slightly earlier than its more famous counterpart. Prepared in Hoffman Mill in Western Australia from at least 1924, it had reached Burnie in Tasmania by late 1926 and Rockhampton in Queensland by early 1927.[3] Within a few years this delicacy was being sold at Grace Brothers in Sydney.[4] Other culinary plays upon the mountain included 'Kosciuskos', a type of cake filled and topped with cream 'like snow', and fruity 'Kosciusko Cakes' were apparently 'very suitable for afternoon tea'.[5]

Unsurprisingly, comparisons with ice cream abounded. In the 1920s Australians could buy a 'Kosciusko Sundae' in Grafton in northern New South Wales or visit the 'Kosciusko Sundae Shop' in Sydney.[6] Through similar such word-play the mountain also entered Australia's commercial and domestic kitchens. 'The refrigerator that gives satisfaction' was, of course, 'The "Kosciusko" Ice Chest.'[7]

In the savoury department Kosciuszko helped sell meat pies. 'From Kosciusko's peaks to north of Cairns', announced one advertisement, 'Rex Camp Pie is always to be had.'[8] Playing on distance and travel and preservation, these tinned pies served a market beyond the ice boxes of Australia and tapped into Kosciuszko's deeper role in a developing national psyche, or at least a growing cultural canon.

Above all, Kosciuszko was the setting for Australia's favourite poem. Banjo Paterson's ballad *The Man from Snowy River* was still only a generation old, itself hearkening back to a time still in living memory:

And down by Kosciusko, where the pine-clad ridges raise
Their torn and rugged battlements on high,
Where the air is clear as crystal, and the white stars fairly
    blaze
At midnight in the cold and frosty sky,
And where around the Overflow the reed-beds sweep and sway

To the breezes, and the rolling plains are wide,

The Man from Snowy River is a household word today,

And the stockmen tell the story of his ride.[9]

In fact, this was no mere poetic licence. Old stockmen did tell stories, and through one of my high-country family branches I can trace a line of storytelling to an ancestor who reportedly inspired Paterson's poem.[10] This rider, Jimmy Kiss, only died in 1921.[11] Having spent some of his later years running hotels in the Monaro region, he very likely knew some of Evan's people, and maybe even Evan himself. Such poetry was wistful, imaginary, but also grounded in the lasting historical importance of Kosciuszko to many Australian families.

Within Paterson's work this comes through clearly. The mythic rider from Snowy River was not his only pointer to Kosciuszko's literary significance. Frequently using the wider region as a poetic setting, Paterson set the humorous 'A Mountain Station' a little downriver of 'Kiandra's snow', for example, and 'Frying Pan's Theology' took place in the Monaro while 'Snowflakes are falling/ Gentle and Slow'.[12] Both belonged to the same collection of 1895 poems headlined by 'a stripling on a small and weedy beast', and again tapped into real experiences.[13] The former focused on cattle-rearing, while the other was a short conversation between an Aboriginal man 'and Boy (on a pony)'. The latter, while reflecting the racial attitudes of the time, provides another curious association with the deeper Monaro past, where Kiss boys – accomplished cattlemen – raced Aboriginal riders.[14]

Paterson once again followed 'Riverina Sheep' up towards Kosciuszko for largely comedic effect in his third collection of poetry, published in 1917. 'The Mountain Squatter' played on annual grazing patterns.[15] But in 'At the Melting of the Snow', he wrote lovingly of his own longing and nostalgia for the high country:

Let us saddle up and go
To the old Monaro country
At the melting of the snow.[16]

With Paterson's three collections published as a single volume in 1921, this poetic imagery of Kosciuszko had a profound effect on Australians as they tuned their collective sense of nationhood through word and rhyme. Even an immigrant like Laurie was likely to be caught up in this.

Yet Kosciuszko stood for more than nostalgia or country idealism. For C. J. Dennis, another of Australia's great poets of this period, Kosciuszko could mean fixedness and constancy. This was played to effect when in 1926 'The continent shook; Kosciusko fell down', and the Commonwealth meteorologist of the day declared it was 'nothing unusual', apparently a contemporary running joke at the meteorologist's expense.[17]

Beyond snow, white-haired old age and steadfastness, Kosciuszko also stood for change. Another poet, the Sydney-based Roderic Quinn, wrote in the 1920s of

Shrill screams the cold south-wester
And, as it wailing goes,
It tells of snowy mountains –
Tall Kosciusko's snows[18]

In Sydney such chilly gales whipped the harbour into a dangerous place, occasionally causing trouble for boats and shipping. Among the harbour's fleet was the little ferry *Kosciusko*, which regularly suffered mishap in the 1920s.[19] The steamboat ferrying Sydneysiders about typified the progress of the age, and linked two of Australia's great icons.

While becoming an essential element of Australian cultural nationalism, the very word Kosciuszko therefore also hinted at a deeper cultural internationalism and transformation. This ambiguity is well captured by two Australian waltzes that bore Kosciuszko's name. The first, composed in the 1910s by F. W. Kidd, was *Dawn on Kosciusko Waltz*.[20] With the music sheet's cover depicting a mountain stream, it was obviously tied to the mountain. But a 'Waltz for Pianoforte' of 1921 – titled *Kosciusko* and dedicated to an Australian politician – was written by the Polish composer Stefan Polotynski with decidedly resonant ambiguity.[21] People might well have wondered if he meant the man or the mountain. Such iconic correspondence between the nations did not escape notice.[22]

More particularly, Polotynski's waltz also reveals some of the cultural connections that drew the worlds of Evan and Laurie together. Copies of the Polish-inspired pianoforte music may well have been handled by Evan as he worked for the Aeolian company in Sydney. And as a former prisoner of the Bolsheviks in Soviet Siberia, the composer Polotynski helped highlight Australia's collective experience of the wider world brought to its shores by people like Laurie.

High Society

Already the type of couple to attract attention in the society pages, the prospect of Laurie and Chrissie marrying went beyond noteworthy to being newsworthy. By December 1923 the engagement was public knowledge in Sydney, even though Chrissie was still overseas, by then staying with the Seamans in Glen Cove.[23] 'Miss Bell is at present spending some time with her fiancé and his father and mother at their Long Island home after having toured through Europe and the British Isles with them', a news report rather casually mentioned.[24]

Large photographs of Laurie and Chrissie dominated the *Sunday Times*'s 'Brides and brides to be' segment in a January 1924 issue.[25] Laurie was the only bloke with his image on the page, partly by way of social introduction and possibly a testament to his quality as a catch. In June, they married in Sydney. The *Sunday Times* reported the event at some length, capturing some of the wider interest in the international couple:

> A charming wedding was celebrated at St. Phillip's, Church
> Hill, recently, when Miss Christiana De Rossi (Topsy) Bell,
> only daughter of Mr. and Mrs. E. W. Bell, Royston, Church-
> street, Burwood, was married to Mr. William Laurie Seaman,
> of Glen Cove, New York ... The bride, who was given away
> by her father, wore a frock of ivory satin meteor, draped on
> classical lines, and fastened with a spray of orange blossoms.
> The train of ivory georgette was mounted on tulle. The
> tulle veil, edged with silver, was held in place with a wide
> bandeau of silver tissue leaves, and a sheaf of white lilies, tied
> with silver tissue, was carried ... The church was artistically
> decorated with palms and large baskets of Autumn leaves. ...
> Among the guests were the Consul-General for America and
> Mrs. Lawton, Sir Charles Rosenthal, Lady Rosenthal, the
> Vice-Consul for America ...[26]

After the wedding, two more photographs of Chrissie featured in July editions of the *Sunday Times*.[27] Captions for both described her as Mrs Laurie Seaman and mentioned she was leaving Australia with her husband for New York. The new couple left in July by a New Zealand–flagged Union liner, the *Maunganui*, which made a regular run between Sydney and San Francisco via intermediate ports like Wellington.[28] But the couple spent less than a year in America, returning to Australia in May 1925 aboard the *Aorangi* via Vancouver and Auckland.[29] They arrived in time for the Kosciuszko season in which Evan was photographed sprawled in the snow.

Laurie started working with his father-in-law, Ernest Walter Bell, who was an established fire assessor and insurance accountant.[30] Laurie became 'controlling officer of the Motor and Engineering Department'.[31] Such work meant regular travel throughout rural New South Wales, which likely appealed to the roving Laurie. The partnership proved sufficiently successful to become formal within a year, and the men ran the business of 'Messrs. E. W. Bell and Seaman' from offices in Pitt Street, Sydney, within walking distance of the Millions Club, with which Laurie was soon associated.[32]

Upon her return to her homeland Chrissie reconnected with the charitable sector, but she also became involved in new initiatives. In September 1925 she met for a 'social afternoon ... devoted to sewing for the Women's Hospital', as part of a women's group called the 'American Circle'.[33] This was a social group for women from America moving to Australia, and marked her new status as a person with international experience. As the Sydney *Sun* explained a month after Chrissie's return:

The first year in a strange country is very hard, and in an
effort to mitigate this strangeness, and permit of friendships
among American women in Sydney, the American Circle came
into being.[34]

Being Australian, Chrissie would have been a useful asset to this
group, which met monthly at a cafe. But the group was more than
social; its members also sewed clothes for children in hospital and
provided toys for the Salvation Army's children's home.[35]

In June 1926 Chrissie's mother died.[36] The cause was 'heart
failure, following on [from] pneumonia and influenza'.[37] The funeral
was so swift that it seems the event might have been expected: dead
one morning, Mrs Bell was buried the next.[38] Chrissie clearly took
after her mother in some respects: only about a week earlier Mrs
Bell had been hosting a table at an event for the Sunbeam Free
Kindergarten in Alexandria.

Chrissie was now a mother herself. In the 'Social Interests'
section of one 1927 edition of the *Sun*, she and young American-
born Bruce Laurie Seaman were pictured at top-centre page.[39] In
fact, a collection of photographs from this period capture the young
family settling into life in Sydney, including building a home at
Palm Beach and enjoying magnificent views from their verandah.[40]

While childrearing obviously kept her busy, Chrissie continued
to be involved in charitably focused social activities, particularly
those connected with children. In April 1927, for instance,
Chrissie organised a table of guests for a dance held to benefit the
Sunbeam Free Kindergarten, where she 'wore a beautiful apricot
georgette frock, heavily beaded in silver, [and] was accompanied by
Mr. Seaman'.[41] The pair had become quite the noticeable couple.[42]

Laurie too settled into Australian society, helped rather than hindered by his American connections. He attended the University of Adelaide's jubilee celebrations as a representative of Swarthmore College.[43] Universities from around the world were represented, including Jerusalem and Singapore, and many with connections to New Zealand, Britain and America. As well as the national and international guests, there were prominent South Australians among the 1300 people in attendance including the premier, government ministers, commissioners, judges and the consul of the United States.[44]

Acting as an international delegate for his alma mater, Laurie mixed with a crowd that included the Antarctic explorer Professor Sir Douglas Mawson, whose Kosciuszko dogs had by now reportedly gone feral.[45] Some had escaped captivity and interbred with dingos and other wild dogs, providing a problem for pastoralists and giving rise to mistaken sightings of Tasmanian Tigers.[46]

By 1927 Laurie was demonstrably connected with the Millions Club, through which he met Evan. In October he was shown in the *Sydney Morning Herald* swinging a golf club for their annual Golf Championship, a sign of his newfound Australian social status.[47] Among other things in the busy year of 1927, Laurie also found time to summit Kosciuszko. New to the sport of skiing, he reportedly 'proved an energetic novice and succeeded in reaching the Summit after a week's experience'.[48]

With that, the new Australian had a new love. And a new friend.

Millions Club

Despite their outward differences, Evan and Laurie shared a passion for Kosciuszko. But what specifically brought them together was the Sydney Millions Club. Individually, they can each be seen undertaking certain club activities, but seemingly not together outside of the mountains. Evan attended the annual Millions Club Ball in 1928, for instance, perhaps his last Sydney dance before departing for Kosciuszko that year, but Laurie was not mentioned as an attendee.[49] Similarly, Laurie was an active Millions Club golfer, but there is no obvious sign that Evan was too.[50] Yet Evan and Laurie were both active members of Sydney's Millions Club, meaning that their paths likely crossed more often than surviving documentation suggests.

Sydney's Millions Club grew from a concept that was developed overseas in the late nineteenth century. Businessmen sought to spruik their cities, and formed clubs intended to encourage population growth beyond certain thresholds. By making these cities attractive places to visit and live, and through bringing people interested in the development of a city together, these clubs aimed to benefit members and citizens alike.

Back in the 1890s Australians had followed the story of the Two Million Club of Chicago with some interest.[51] It aimed to beat New York in reaching two million inhabitants. By the early twentieth century a visitor to Sydney from Chicago, Robert C. Givins, inspired talk of an equivalent Australian club.[52]

Givins's first impressions of Sydney were highly favourable. When his ship approached Sydney Heads, someone banged on his cabin door. It was about 5 am, and he had been asleep. 'Say!' a voice called out. 'If you want to see the finest harbour in the world come on deck!' Heeding this call, Givins made his way to the deck and was duly impressed, later describing it for his fellow Americans in a report of his journey:

As the ship ploughed through the smoother water up the
bay I agreed with my friend, for before us opened the most
magnificent panorama imaginable. The various inlets and
coves, together with the harbors and the lower portion of the
Parramatta River furnish sufficiently deep water to anchor the
entire naval fleet of the world … I cannot remember having
seen anything quite so picturesque as Sydney Harbor.[53]

He then turned his attention to the city, and saw in it similarities
with Chicago:

The city is building up and improving in every direction
and especially in the suburbs, and new tracts of land are
being placed on the market similar to the Chicago method
of developing suburban towns. The houses and buildings are
brick and stone, and the city, built on high ground, shows
off to great advantage. On the streets, as in Melbourne, one
witnesses the same styles in dress and clothing as in Chicago.
The city is progressive and its citizens patriotic, every one
doing the best he can for Sydney.[54]

During his journey, the American bonded with his hosts, and noted
that 'all our jokes and stories are in active circulation here'. He was
reportedly hushed by some women when he tried to tell them one –
they had already heard it some years earlier. The anecdote was a
sign for his readers of the strong cultural links between these nations
on either side of the Pacific Ocean. He saw a great future in the
two nations working together. Apparently when he told a couple of
Sydneysiders about the Two Million Club someone sprang up, saying,
'Capital idea! We will establish a One Million Club in Sydney.'

It was a few more years before the idea came to anything. But following a study tour of America in 1911 the director of the New South Wales Immigration and Tourist Bureau publicly suggested that Sydney would do well to have its own Million Club, which could focus on 'improving the city and all its services and conditions of life, so that Sydney should stand as the hall mark of perfection, and attract visitors and population from abroad'.[55] On 5 January 1912, the Millions Club was formed at a meeting at the Australia Hotel in Sydney.[56] The organisation swiftly became a major social and political hub for Sydney. By 1926 Evan was attending their annual dance.[57]

Because Millions Club functions often attracted press attention, such reports give a rich history of political, economic and scientific discussions within Sydney. Apart from purely social events like dancing and golf, lectures and talks often focused on questions of progress, whether social or technological. In the early part of the winter of 1928, for instance, shortly before Evan and Laurie left Sydney for that ski season, the famed aviators Charles Kingsford Smith and Charles Ulm addressed a Millions Club audience.[58] Having recently completed the first trans-Pacific flight from America to Australia, the aviators gave a presentation about their adventure. Smith apparently thought 'that the flight would, in all probability, still further cement the feelings of friendship existing between Australia and the United States'. Having borrowed its core concept from America, the Millions Club also fostered a cultural and intellectual rapport between the two countries. It was the right environment for Evan and Laurie to become friends.

Being American himself, and a former military aviator, Laurie might have been particularly interested in the Smith and Ulm event. He certainly demonstrated some enthusiasm for the airmen's feat because that month he contributed to a fund being raised for them.[59] Leaving Sydney early in August these aviators flew in the *Southern Cross* to Melbourne, where they were quizzed by reporters about their flight.[60] 'We were very cold,' Smith told the *Age*'s journalist, 'especially when flying over Kosciusko, even though we were only 6000 feet up. With the sun glittering on the snow the mount presented a sight long to be remembered ... the spectacle – well, it was worth it.'[61]

Far below, Laurie was supposed to be heading towards Kosciuszko himself. He too had a penchant for fast mechanical travel.

Speeding towards Kosciuszko

As well as being a Millions Club member, Laurie was active in the Royal Automobile Club of Australia. Surviving Seaman family photographs contain several images of motor cars, highlighting how they were a fixture of Laurie's life and Chrissie's too. One shows Chrissie shovelling sand from around a bogged car, which was tied to a horse preparing to tow it free. Another has Chrissie seated inside, with Laurie standing by. Motoring was clearly of great interest to the family and part of their everyday life. But Laurie's involvement was more than a leisurely pursuit.

In March 1928, for instance, Laurie participated in 'an interesting contest' that pitted cars in a testing run from Sydney to Avon Dam, about 100 kilometres south of the city.[62] Laurie's entry was a six-cylinder American Roadster, the 3127cc Buick Six. While competitive, it was more of an engineering trial than a race. Participants were comparing their vehicles for fuel efficiency, reliability and various other technical capacities.

The contest had several elements. Drivers and cars would meet in the city at 7.45 am, bringing a specified length and width of rubber tubing, which would connect a 'special petrol container with the carburettor'. This way each car ran on precisely the same volume of petrol for a fuel efficiency test during part of the day. Formally starting from a weighbridge, where weight and departure times were logged, the drivers were advised that their 'speed from that point will be moderate to avoid difficulties with the traffic'.[63]

The drivers made their way through the outskirts of Sydney, Liverpool and Campbelltown and into the country.[64] The twenty-six other vehicles in the contest included a 748cc Austin Sports, a 1496cc Bugatti Four, and cars by Vauxhall, Cooper, Chrysler, Studebaker and several other manufacturers. At Narellan the fuel consumption test commenced: the special fuel tanks were connected by officials, and the drivers sent on. As they now had to climb over Razorback Ridge, this tested the engines on difficult roads. Laurie made good progress up the hills, but his special tank ran dry 'within 50 yards of the crest of Razorback', which was considered 'most unfortunate', as he could have coasted home if only he had beat the crest.

By this point seven other cars were also out of the running, two of them driven by women: driving was not exclusively a male domain. Yet despite being out of contention for the fuel consumption prize, the drivers continued to the dam, and Laurie did better in 'the flying half-mile'. This took place 'on the road inside the water reserve boundary near the Avon Dam, an excellent course with a hard, smooth tarred metal surface, slightly undulating, but otherwise quite straight.' Here the cars

were pushed for top speed, and the fastest managed to average
79–80 miles per hour, completing the course in a fraction over
22 seconds. Laurie's Buick came a respectable fourteenth, in the
middle of the pack at 32 seconds.

Laurie took his Buick on another testing run to Avon Dam in
April, where the organisers provided a greater fuel allowance for
the run over Razorback Ridge.[65] On the final straight towards the
dam, Laurie averaged 58½ miles per hour, but was disqualified
from the overall competition because he 'fell foul of the officials'.[66]

In May the Royal Automobile Club tried something different.
Their next event was a distance trial, aiming to cover a sweep of
road from Sydney to Moss Vale, Kangaroo Valley, Nowra, Moruya,
Goulburn and back to the clubhouse, all within twenty-four
hours.[67] Laurie was among twenty-nine competitors on this journey
that would take him to the Southern Highlands, along the coast
and traverse the Great Dividing Range twice.

The event started smoothly, and the drivers made good time to
the first descent, where they were slowed to an average speed of
only 20 miles per hour.[68] All parties reached Nowra safely, where
the local hotelier 'supplied all competitors and participants with
hot coffee and refreshments in the middle of the night when they
arrived'. Although there were no women drivers in this competition,
there were women and children among the passengers.

Most of the cars were enclosed, but some of the drivers of open
vehicles had to wrap themselves in rugs against the reportedly
bitterly cold weather. The night driving surely exacerbated the
difficulties presented by these conditions. The darkness also caused
problems for one driver, complicating a roadside repair to their
petrol tank, which was punctured in the dark:

A stone thrown up by one of the wheels struck the tank, and made a hole, which was plugged up as well as possible with pieces of wood.[69]

Incredibly, the car continued. After refuelling, it 'left with a strong stream of petrol pouring out of the puncture', but still completed the full circuit.

At Bateman's Bay, the drivers crossed the Clyde River in a punt, and ran down to another river town, Moruya. From there they turned inland again, climbing up the Araluen Valley back over the ranges to Braidwood. As they left the coast behind, they returned to the worst of the late autumn weather. Around Braidwood, the drivers struggled against gale winds and a reported 'sandstorm', but the relatively flat plains meant they could make up for lost time.[70] In rounding a corner a small Austin 'turned turtle', throwing a passenger as it rolled. Luckily the passenger survived and was picked up by another driver for a safer journey back.[71] The driver of the overturned Austin, however, simply righted his vehicle and kept going. In the final run, as the cars passed through Picton, one of them hit a pig. 'The machine and the animal escaped injury', one report quipped.[72] With that last hiccup the convoy raced towards the finish line at the clubhouse.

In the end, the May competition resulted in a split victory. Having earned 500 points for his journey, Laurie shared winning honours with twenty-three other drivers.[73] He may have had some close encounters and difficult corners, but no misfortune spectacular enough for press attention. Punctured petrol tanks and overturned cars made more impressive stories.

In this age of road-testing and car-contests Laurie was doubtless as enamoured of the adventure and thrill of it all as the other competitors. But the testing also had serious engineering and marketing implications. Because of the blustery conditions, commentators gave much attention to the performance of the vehicles in cold weather. While defending open cars for their engine warmth (the engine acted as a sort of heater for passengers), the *Sydney Morning Herald* argued that the enclosed cars 'gave a fine demonstration of the suitable character of this type of body to Australian touring conditions under such weather conditions'.[74]

This observation was significant, because cars were increasingly touring up towards Kosciuszko. The Royal Automobile Club was even planning to conduct one of its reliability and touring contests on 18 and 19 August 1928 as part of their club's skiing holiday.[75] They planned to terminate the event at Cooma, where the cars would be garaged while the club members were taken up to the Kosciusko Hotel in more suitable vehicles.[76] Laurie was singled out in the advance publicity as a serious contender for the club's skiing championship but he was not one of the listed participants for the driving and touring part of the trip.[77] By the time the cars left the city he was already in the mountains with Evan and the Millions Ski Club. Like thousands of other Australians, they took the Kosciuszko train.

The Kosciuszko Train

At precisely 8.20 pm on 10 August, 120 Millions Ski Club members left Sydney on a 'special train … for the seventh annual carnival at Hotel Kosciusko'.[78] They were due to return to Sydney eight days later.[79] Chrissie did not join Laurie for the trip, staying home with their young son.

Technically, it was the Cooma Mail train. Connecting Sydney to Australia's Alps, the final section of railway was completed in

May 1889.[80] Crossing creeks and ranges in difficult terrain, the line reached over 822 metres above sea level at its highest point. Although various proposals to extend the railway to Jindabyne or even Kosciuszko had been proposed over the years, and branch lines reached elsewhere into the Monaro, in the 1920s Cooma was still the terminus point for the first stage of a journey to Kosciuszko.[81] Many Australians thereby experienced much the same journey into the mountains, making it one of Australia's most famous rail journeys.

Millions Club skiers had travelled to Kosciuszko by rail since at least 1925. The club benefitted from an arrangement between the Government Tourist Bureau and the Railway Department to the tune of 'five cars of sleeping births' attached to the Cooma mail train.[82] While it may sound like club members must have opened a few public service doors, this was a mutually beneficial arrangement as it both promoted and facilitated Kosciuszko tourism and domestic leisure travel more broadly. Special Millions Ski Club trains followed in the winters of 1926, 1927 and 1928.[83] The train journey was clearly considered a significant part of the holiday, a ritual break from suburban normality.

The Tourist Bureau was actively supporting train travel to the snow, especially for such clubs, but also for schools. In fact, much can be gleaned about the experience of the Kosciuszko train from school excursions into the mountains, which often resulted in many small essays about the journey. Regularly published in local newspapers, they help document the winter seasons in the Snowy Mountains and the sights and events along the way, especially the train trip.

In July 1926, for example, fifteen-year-old Rod Mackay travelled from Sydney on the Cooma Mail train.[84] He recounted boarding in the evening and putting 'our luggage in the racks', before turning to the journey itself:

> As we moved out there was a hullabaloo of shouts, war cries, and the tooting of other engines. It was a great send-off, and left us in a very happy mood. We were too excited and noisy to sleep. When the moon rose several of us watched the countryside.

For city children like Rod, such journeys obviously helped them to develop an affection for the countryside. It is easy to appreciate how a bright winter moon rising above the bush and the plains could be a compelling feature of such a train journey, especially as the carriages rattled into the wide open Southern Highlands.

An unnamed 'scholar from Gosford' detailed their journey in June 1928, just a few weeks before Evan and Laurie followed the same path:

> The night was clear and not too cold for travelling. Very little sleep, as many expected, could be had; the voices of excited pupils being audible throughout the night. At Moss Vale refreshments were partaken of, and some confusion arose getting back into the train. At Goulburn, many were looking out for the memorial which showed out very clearly against the black sky.[85]

The tower on Rocky Hill just outside Goulburn was one of New South Wales's most notable new war memorials. Unveiled in late 1925 it was, of course, part of a larger phenomenon in Australia during the 1920s of commemorating the dead of the Great War.[86] Focal points for communal grief and commemoration, war memorials across Australia bore the names of dead sons, brothers and fathers, many of whose graves the locals could never hope to visit. Such places often became sites of interest for visitors, and the Goulburn tower was certainly an object of such tourism. The hill on which the tower was built had previously been proposed as a good situation for a tourist drive. By the winter of 1928, when the Kosciuszko train of Gosford schoolchildren was chugging by, regular weekend tours were being managed by a local caretaker.[87]

Other young writers penned similar thoughts, including Bessie and Eileen Clancy, who travelled with a group from Grafton High School in late June 1928, about a week after the Gosford children.[88] Like other country kids, the Clancy girls had spent many hours on the train before they even reached Sydney. They too spoke of 'war cries' as the trains pulled out from Central Station, as well as 'school songs and the cheering of over one hundred happy school children from the High Schools of the Northern Rivers'. Such correspondences between accounts of children travelling with different groups on different days may point to a sort of departure tradition. Perhaps the stationmasters or railway staff encouraged the hubbub. Train travel itself was a noisy phenomenon, so a little more may not have hurt. But, as Bessie and Eileen suggested, it continued beyond the station:

> To those among us who enjoy the train travelling the journey to 'Kossy' was a piece of fun. As the train rattled on snatches of song, war-cries and the buzz of conversation mingled with the scream of the whistle, disturbing the moonlit country with a strange medley of sound.

The next point of interest for many young correspondents was breakfast at Cooma. As the Gosford scholar wrote, 'the cold was felt very keenly, mainly attacking the feet, which, however, soon warmed in walking to the hotel'.[89]

Cooma's weather seems to have greatly affected some visitors: it was 'a very cold place', reported one contingent of schoolgirls from Warwick in Queensland, who also travelled up that winter.[90] And while most commented on there being a hotel breakfast, few gave many culinary particulars. Perhaps because it was the place where the train terminated most writers simply moved on to the next part of the journey. The Gosford scholar broke with this convention by reporting how immediately after breakfast some children took a short walk around the town, 'which, although appearing old, was very pretty'.[91] For Evan, the town would also have been deeply familiar, in every sense of that word.

School groups boarded smaller vehicles at Cooma for the rest of the trip to the Kosciusko Hotel. These vehicles were variously described as cars, coaches, lorries and so on. Like the train, the convoy departed with 'war cries and cheers' – at least, so reported one child that winter.[92] 'The sun shone brightly, and was very welcome after the cold frost in Cooma', the Gosford scholar noted in July, but this report was an anomaly.[93] The Clancy girls from northern New South Wales characterised this stage as 'one vain endeavour, on our part to gain protection from the cutting wind'. Another young correspondent, who travelled earlier in the year, also mentioned how from Cooma 'owing to the cold wind we were forced to take more care for ourselves than for the scenery'.[94] But the destination always seemed to be worth the trouble.

The convoys passed through Berridale, sometimes stopping for a 'refreshment' among the small town of 'about two stores and half a dozen houses'.[95] Travelling groups mostly seemed to rest at Jindabyne. 'There we got out and some of our party had a game of "footie" with the local lads by the Snowy River', declared one writer rather nonchalantly. The Gosford scholar mentioned that the Snowy River was partly frozen, and that 'pebbles were gathered for mementos' from the famous waterway.

Each group strained for their first sight of snow, which was generally one of the high points of the travel narratives. Young Rod Mackay captured the general sensation well:

A few miles from Jindabyne snow-covered peaks could
be seen in the distance. The snow began to appear about
the country around us. It was deposited on the tree leaves
and branches, and looked like beautiful white wool. The
fall collected in especially large quantities on old tree
trunks.[96]

The sight of snow, unsurprisingly, also tended to lead to more noise. 'As the patches grew bigger our eyes grew brighter and at length our excitement found vent in wild "hurrahs!"', said the Clancy girls. As another of the young travellers of 1928 wrote:

It was a sight never to be forgotten, snow-tinted mountains,
with the sun shining on them, made the snow appear like gold
sprinkled among dark trees! Then we began to ascend the
mountains, and soon were among real snow! Everything was
white, and the sun made it glaring in our eyes, but what did
we care? Looking down into the valleys, now and again, small

farm houses could be seen, and these made one wish that I
could live there forever![97]

This same writer concluded the journey with his arrival:

Eventually we came into sight of the hotel. Down in the snow-
clad golf links we could see happy, care-free people enjoying
their 'spills' and 'fights' in the snow. Oh! Whoever has seen it,
will they ever forget?

His letter ended there in front of the Kosciusko Hotel, but fortunately
others kept writing, revealing the hotel in all its charm.

Arriving at the Kosciusko Hotel

By 1928, after less than two decades of operation, the red-roofed
Kosciusko Hotel had become one of the most famous symbols of
modern Australia. It had survived near-misses with bushfires and
political pressure to close it during financial troubles.[98] Blending the
look and feel of an Australian hotel with a European chalet, the
Kosciusko Hotel – like the journey to get there – was a unique and
memorable experience.

Even the well-travelled appreciated it. After a much-publicised
visit to Papua in 1926, for instance, Adelaide journalist Effie
Sandery visited the Snowy Mountains.[99] She described 'stiffly
moving passengers climbing from cars, and entering the hall which
always seems to be filled with young folk talking and standing
with their backs to the log fire.'[100] Her article about Kosciuszko,
intended for a far-flung readership in generally snowless South
Australia, bore the subtitle 'Hilarious sport of ski-ing'. But while
mirth was an important theme, for the most part the story was one

of enchantment. Sandery's version of arriving at the hotel focused on the contrasting colours and textures of the people:

> Figures, dark, muffled, scarved, booted, and gloved, pass by with skis either on their backs or on their strangely behaving feet, the bright colours of jumpers and other clothing standing out vividly as they approach.[101]

Even while describing the interior of the hotel, she swiftly shifted to the subject of clothing:

> Small, comfortable rooms looking out over more snow; a hasty change of clothing; and down again to the ground floor, to the ski rooms where the skis and boots and puttees which are so necessary may be hired by the day or week. There are few skirts to be seen at Kosciusko, except in the evenings, when out of three layer[s] of woollies the female sex climb into two layers of crepe de chine.[102]

As Sandery so clearly articulated, the Kosciusko Hotel was a place of relaxing charm and considerably gaiety.

Laurie's impressions went unrecorded, but upon arriving at the Kosciusko Hotel in August 1928 Evan quipped that he 'had practically to pawn his shirt' to make the trip.[103]

Evan's comment mixed his love of the place with a joke at his own expense. Skiing was certainly gaining widespread popularity as a holiday activity, but regular enthusiasts tended to be wealthy professionals like Laurie who were better able to afford the travel, accommodation and equipment required. While the government was encouraging Australians to holiday in the Snowy Mountains, such excursions did not come cheaply.

The schoolchildren therefore reflect something of an aspirational equality, where all Australians could have the opportunity to experience the mountains, but they were also an investment in the future of alpine tourism. Which is just as well, for they again provide the richest accounts of the experience.

Arriving a few weeks before the Millions Ski Club, for instance, the Clancy girls described the Kosciusko Hotel as 'a cosy, red-roofed building at the foot of a hill. Snow-capped peaks surround it, a stretch of dazzling white extends in front, with a rough circle of ice (which is the frozen lake) in the centre of this whiteness.'[104] They also referred to being 'welcomed with a hearty snowballing from children of other parties who had arrived there before us'.

The children were generally fed upon entering the hotel. Then they were given 'a theoretical lesson in ski-riding' and set upon the nearby slopes.[105] But they were not the only ones who were learning to ski. Apart from the foreign-born or those raised in the high country, skiing and ice-skating were relatively new sports for most Australians. Indeed, they were relatively new for the world, highlighted by the fact that the first Winter Olympics was only held in 1924.

Because the visiting children and their readers were generally unfamiliar with the sport, descriptions of learning to ski were a common part of the snowbound school narratives. Some children from Armidale and Brisbane, visiting about the same time as the Clancy girls, started their joint account of the sport by describing the equipment:

The skis are made of wood, four inches wide, and ranging
from about 4ft. to 10ft. in length with an upturned and

sharpened point. A toe-iron, toe-strap, and a strap buckling round the heel of the boot keep them firmly in position and as soon as one gets one's balance, the sensation of at first gliding, and gradually gaining pace down the hillside is a most exhilarating one.[106]

Most such stories then turned to rounds of falls and bruises and yet more attempts, as the ski fields became learning curves for the newly initiated. The Clancy girls capture this delicate interplay of experience, confidence and hubris well:

With a good deal of practice and many falls we grew accustomed to the tricks of the ski and then there was nothing so heavenly as skimming down a slope with the snow flying behind, a sharp bend looming ever nearer and the question throbbing 'Shall I get round it?' Alas, as soon as one's thoughts are distracted one sits down heavily on the track – a wonderful tangle of arms, legs, stocks, snow and ski.[107]

'They had many falls, much fun,' one teacher said of her students that winter, 'and felt fit and happy.'[108]

The adventures even extended into the evening, as the Clancy girls related:

At five o'clock each day our fun on the snow had to end, and, leaving our heavy boots and breeches to dry in the drying-room, we donned our daintiest and lightest attire for dinner and the dance which followed in the ballroom of the Hotel. This was a magnificent ballroom with steam radiators placed at regular intervals around the walls, which made the room delightfully warm, as was, indeed, every other room in the Hotel. To the strains of the Hotel orchestra we tripped the light fantastic until eleven o'clock each evening and on occasion of a fancy-dress ball until 12.30.[109]

Other child writers frequently mentioned the fun, fancy balls, the library, picture shows and various games that occupied the visitors during the winter nights at the Kosciusko Hotel. One of the teachers referred to boys 'jazzing merrily' into the night.[110] Another teacher estimated the hotel was accommodating some 280 people in the middle of winter 1928, about 260 of whom were schoolchildren.[111]

Like most great hotels, the Kosciusko Hotel was therefore itself an attraction, drawing guests from around the globe. But after arriving in August 1928, neither Laurie nor Evan stayed there for long. As Evan told a friend, 'I am going out amongst the big Alps stuff. I am not going to stay about the hotel.'[112] On Monday 13 August, in company with several other friends, he and Laurie abandoned their comfortable lodgings and headed higher into the mountains towards Kosciuszko.

CHAPTER FOUR

Stormclouds

Australians still had much to learn about Kosciuszko when Evan and Laurie left the hotel for Betts Camp in the winter of 1928. It had, after all, only been 31 years since the first recorded winter summitting of the mountain. Certainly, knowledge had advanced greatly during Evan and Laurie's lives, but even still Kosciuszko remained a geological frontier, because the main feature of Australia's Snowy Mountains is not snow, but rock.

Geological expeditions helped scientists refine their understandings of the deep history of Australia, and accounts of their journeys helped spread their science. One expedition from the University of Sydney in early 1922 drove to Kosciuszko's summit, gave three cheers 'for Kosciusko and for its discoverer', and then carefully examined the signs of ancient glacial action upon the landscape.[1] From alpine lakes to rocks carved by the work of ice, these geologists built up a picture of Kosciuszko as it once was. 'During the maximum glaciation,' they suggested, 'the Kosciusko Plateau must have been a miniature Antarctica', leading then into a sort of scientific poetry:

With the eye of faith one can still see the deeply crevassed
glacier ice, descending from the highest peaks, the long
lines of dark moraine travelling down the glaciers to where
they end in steep ice slopes, undermined by turquoise blue
ice tunnels, whence the sub-glacial thaw water rushes forth
foaming and roaring over the rocks and boulders as it plunges
into the valley below, and one can hear the thunder of the
avalanche and the boom of the cracking ice field.[2]

Here was proof of Australia's icy summit as it was in the Pleistocene,
revealing a continental history that looked and felt very different
from the seemingly eternally dry world of the desert interior.
Kosciuszko's rocks were proof of climate change, its lakes the last
puddles of the ice ages.

Teachers who took their students to Kosciuszko were likely
aware of the landscape's instructive potential. One account of a
trip to Kosciuszko published in a journal for public school teachers
practically adopted a theological tone: 'Besides the scenic beauties,'
the teacher wrote, 'we were constantly confronted with some
geological wonder which ended in veneration when we at last stood
on the top of one of the oldest spots in the world.'[3]

Yet they blended this lithic veneration with an awareness of the
human history of the landscape, noting the 'occasional ruin of a
shepherd's house' along their way. That there was an even longer
history of human engagement with mountain and rocks is attested by
at least one old prospector, who reported finding 'stone tomahawks
of a hard, blue or green diorite on the tops of Kosciusko, Buffalo,
and Bogong' mountains.[4] Between huts and tools and a road to the
very summit, Kosciuszko was evidently far from untouched, but it

nonetheless helped Australians reach much further back in time, into and before the dawn of human history in Australia.

Kosciuszko also helped Australians understand the science of their continent in another, more immediate way. In doing so, it advanced scientific understandings of the workings of global climate.

Inspired by an international meteorological conference in Paris, colonial meteorologist Clement Wragge had established a weather observation station on Mount Kosciuszko back in 1897.[5] Seeking to capture simultaneous observations from a series of key stations, including Mount Wellington in Hobart and Sale in eastern Victoria, Wragge aimed to capture experimental data that would help interpret weather patterns and improve predictions.

Over the next few years the Kosciuszko Observatory occupied the summit, its attendants enduring extreme conditions in Arctic tents and then a wooden hut, taking multiple daily recordings of the mountain weather through every season. It was a remarkable scientific endeavour. Gales destroyed tents, observers got lost and had near-misses, and the observation team was occasionally snowed in, yet for a few years Kosciuszko was called home by a small band of men, one cat, and a St Bernard pup.[6]

Abandoned in the early twentieth century for lack of government support, the Kosciuszko Observatory soon became just another decaying hut in the alpine landscape, occasionally mentioned by visitors to the mountain. One group of geology students summiting Kosciuszko in 1907, for instance, noticed 'the Premier's breakfast menu card' pinned to the dilapidated structure.[7] Plans to reuse the old observatory for tourism came to nothing, and soon enough the hut was a fading memory.

Interest in the science of Kosciuszko, however, had struck a popular chord despite the government's neglect. Lantern shows and lectures about Kosciuszko were popular in this early and heroic age of Kosciuszko science.[8] These brought the scenery and science of Kosciuszko to the public, but also the thrill of winter expeditions to the top of Australia's Snowy Mountains. Like the contemporary Antarctic expeditions, the Kosciuszko scientists attracted fame and attention, and as the sport of cross-country skiing grew in popularity and its adherents in confidence, adventurous young men like Evan and Laurie increasingly sought their own expeditions beyond the comforts of the Kosciusko Hotel.

Skiing to Betts Camp

Anyone who sought to ski among the main peaks and valleys of Kosciuszko generally had to camp further afield than the hotel. The small hut at Betts Camp roughly halfway between the hotel and the mountain offered the perfect base. Situated in a valley on the gravel road to Kosciuszko, it provided the last available beds and fireplace before the mountain. In summer, visitors could nonchalantly stop and picnic at the hut on their drive to the mountain, but in winter even accessing Betts Camp itself was an achievement. The road was generally buried by snow, and often enough the hut was too.

The hut was named for surveyor Arthur Betts, who had mapped the Kosciuszko area in the second half of the nineteenth century. One visitor described it in some detail after staying there in 1906:

> Betts' Camp is a weatherboard cottage of two rooms, erected
> by the Government. One of these is a sleeping apartment
> for ladies, and the other for men, the latter serving also for

dining and smoking room. Furniture there is none, save for a few empty cases for seats and a roughly improvised table. The beds are heaps of gum bushes, made softer by the addition of some tussock grass; three coarse gray blankets are provided by the guide, and your pillow is whatever you can lay your hand on. Being the only visitor I took the ladies' room, influenced by the natural conclusion that the bed there would be the more luxurious of the two. After a night's experience I was convinced that those responsible for the sleeping arrangements at Betts' Camp act upon the modern idea of the equality of the sexes.[9]

By 1928 a larger replacement hut had been erected, roughly twice as large, but it was still simple camping accommodation.[10]

Besides shelter, Betts Camp offered a destination for the adventurous. In July 1928, for instance, only a few weeks before Evan and Laurie set off from the hotel, a party of twelve high school boys were guided there by an experienced headmaster.[11] Only the fittest were allowed along, after proving themselves capable on skis. Fortunately, they had perfect weather and the trip only took three hours of cross-country skiing. They made a point of going a fraction beyond the hut to claim the record from the students who had undertaken the trip in 1927 – the first such school group to have attempted the run – and returned to the hotel that day.

The eight-mile run between Betts Camp and the Kosciusko Hotel provided a suitable challenge for competitive cross-country skiing, making it one of the great measures of alpine skill. Experts could manage runs from the hut to the hotel at about twice the speed of the schoolboys.

Like the hotel, Betts Camp was something of an alpine icon. It was a place of some rough familiarity, visited by some Australians in the winter months, by many in the summer, and known of by even more. Yet it also marked a special adventure by being the usual overnighting spot for winter attempts to reach Kosciuszko, and thus became something of a clubhouse. As one report of the 1920s put it, albeit with some exaggeration, 'through the season Betts' Camp resounds continually to the merry voices of those who are contemplating the climb on the next day'.[12] The truth was the hut was often empty, but it did tend to bring together like-minded adventurers.

Evan and Laurie reached Betts Camp on Monday 13 August 1928 in company with a larger party from the Kosciusko Hotel. After reaching Betts Camp, about half the party returned to the hotel. A small group opted to stay the night at Betts Camp, reportedly planning to ski among the higher slopes in practice for the club's forthcoming competitions.

Staying overnight with Evan and Laurie were four other men, united by a passion for cross-country skiing despite coming from diverse backgrounds.

There was Lorne Douglas, a former soldier who had served with the AIF in France and lived in the Sydney harbourside suburb of Mosman in the early 1920s.[13] He was too ordinary a bloke to attract much attention from the newspapers in his day. Another was a Norwegian expert skier named Larsen, a now elusive but once familiar personality among the competitive winter sports community.[14] Larsen was one of two Norwegians who resided at the Kosciusko Hotel as 'experts' during the skiing season.[15] He likely joined the expedition as a sort of guide.

The youngest member of the group seems to have been Tasman Cedric Bottrell, who had just turned twenty-three and was only a few years away from inheriting a fortune. By the terms of his late father's will, Bottrell was due to inherit a share in a trust that was worth a staggering £77, 617 when the will was proved in 1913.[16]

The last member of the group was, like Laurie, relatively well-known. Emil Sodersteen was a few weeks shy of his twenty-ninth birthday, and already quite famous around Australia.[17] The Balmain-born son of a Swedish mariner, Sodersteen studied architecture at the University of Sydney, rapidly rose to prominence in his chosen profession, and already had his own practice in Sydney. As well as designing many of Sydney's office buildings, he had helped design Brisbane's City Hall, and was active in the New South Wales Institute of Architects. Only a few months prior to this ski trip he had gained a prominent role on the national architectural stage: his design for the Australian War Memorial in Canberra was considered the best proposal, and formed the basis for the subsequent formal design, jointly executed by Sodersteen and another architect.[18]

As evening fell at Betts Camp, these six companions would have shared a simple meal from the hut's provisions and spoken of their plans for the next day. They probably checked their equipment, Laurie applying oil to the underside of his personally monogrammed skis. Meanwhile, smoke from their fire would have drifted up the chimney and dispersed in the night sky. Nobody else was outside to see it, for these men marked the very edge of human settlement. In fact, their presence might have gone entirely unrecorded, were it not for what followed the dawn.

Clear Blue Sky

There are different versions of what happened that day, but they vary in only minor details.[19] The basics are as clear as the blue skies under which they journeyed. After breakfasting at Betts Camp, the six men skied west and south over icy snow under a winter sun towards Charlotte Pass.

Cutting the easiest direct route through the ridges guarding the approach to Mount Kosciuszko, Charlotte Pass was a crucial landmark. Through it, or rather over it, lay the Snowy River's uppermost valley, slowly cutting its way down the mountains as it flowed from Kosciuszko.

Named for Charlotte Adams, who in 1881 became the first woman of European descent to record a summit of Kosciuszko, Charlotte Pass marked a crucial waypoint on the main route to the mountain. The road traversed the pass and then shadowed the river, eventually crossed it, and then rose to Rawson Pass, nestled between Etheridge Ridge and Mount Kosciuszko. But in winter, the road mattered little.

For skiers, Charlotte Pass was the entry to that 'big Alps stuff' of which Evan had spoken at the hotel. Besides Kosciuszko, there was an assortment of peaks and ridges. As well as the Snowy River, various tributaries cut the landscape. There were also the alpine lakes like Lake Cootapatamba, immediately south of Mount Kosciuszko, or the Blue Lake, found in quite a different direction immediately across the Snowy River to the north of Charlotte Pass. In that respect, Charlotte Pass was the point at which parties had to make decisions.

Evan and Laurie reached it first. While skiing from Betts Camp they got ahead of the rest of their party. That the twenty-nine-year-old Evan would take the lead was unsurprising. He was, after all, an accomplished athlete on Australia's ski fields. That Laurie kept up with him makes sense too: the thirty-four-year-old had a taste for speed and adventure.

But the others were not far behind. In fact, they were close enough to see Evan and Laurie ascend the rise to Charlotte Pass.

Before making their own climb, Douglas, Larsen, Bottrell and Sodersteen stopped for a break of oranges. They rested beside 'the small red experimental hut' below Charlotte Pass. This was a weather station, more like an equipment shed than the sleeping hut back at Betts Camp, but it was the only human-made landmark worth the name in the immediate area considering the road was buried under snow.

Looking up at the ridge above the little red hut, these four could see Evan and Laurie make some sort of a gesture then turn and ski on through the pass.

Towards Kosciuszko

The group's original plan had simply been to reach Charlotte Pass. There had been some discussion about the possibility of heading towards Mount Kosciuszko itself once through the pass, but nothing firm. Despite the gesticulations from the pass, the four men who followed Evan and Laurie were unsure exactly where their friends had gone.[20]

Icy snow meant that it was possible to ski without leaving much of a mark, so even the ground was no help. There was no obvious trail to follow. If they were heading towards Kosciuszko itself, then they were already out of sight, deep in the Snowy River Valley that curved towards Australia's highest peak. Another possibility was that Evan and Laurie had turned right, intending to loop back to Betts Camp via a stretch of the Snowy River and another pass by the Perisher Range.

While perplexed, they were unconcerned. The whole group had vague plans for a late lunch back at Betts Camp before skiing back to the Kosciusko Hotel for an evening dinner and club meeting. Their morning excursion was meant to be a short one because the return trip was long. Douglas, Larsen, Bottrell and Sodersteen simply assumed they would meet Evan and Laurie back at Betts Camp during the afternoon if they did not otherwise bump into them.

Then the four discovered some tracks in the snow. Pointing towards the mountain, the tracks may have encouraged them to go on and try their luck. Although a journey to the summit would likely throw out any chance of reaching the hotel for dinner, the weather was still good so they started skiing towards the mountain. At one point the group 'saw two figures and made towards them, only to discover that they were two perpendicular rocks'.[21] They were obviously still keeping an eye out for Evan and Laurie.

Their journey paralleled the Snowy River as they followed its valley for several miles. Eventually their path converged with that of the river at the crossing below Etheridge Ridge. This range was the last obstacle before the final ascent. From Rawson Pass, which separated Etheridge Ridge from Mount Kosciuszko itself, the final ascent really began. But while they were somewhere in this area the weather turned. The day suddenly became dark and cold and dangerous.

A Kosciuszko Storm

As the meteorologists had learned a generation previously, Kosciuszko's climate could be fickle. When the Millions Ski Club arrived at the Kosciusko Hotel about midday on Saturday 11 August 1928 the weather had been fine.[22] In fact it had been relatively fine for much of the season. In late July, the weather in New South Wales had been so mild that the Sydney *Sun* asked, 'Is Spring Coming?'[23]

Over the preceding several weeks twin high-pressure systems had dominated the weather of south-eastern Australia. A high north of Adelaide and another off the east Victorian coast kept conditions steady over much of the continent.[24] Temperatures were cold and there was some rain throughout New South Wales.[25]

But the clouds that loomed over the horizon at Kosciuszko that day were part of a larger system bringing change. While Evan and Laurie and their friends travelled to the mountains, then skied to Betts Camp, low systems in the Southern Ocean pushed into the Great Australian Bight, bringing troublesome weather north.[26] Farmers reported that 'Welcome showers fell' at Dookie, and 'a good general rain' soaked Nagambie. 'Steady rain commenced' at Seymour, and it was 'still raining' at Sale as the Tuesday report was wired to the meteorological office. Violet Town's message typified the response of the rural farming communities to the precipitation: 'Rain fell all to-day. Much anxiety has been removed.'[27]

While bringing agricultural relief, this system also brought destruction. During the morning as the skiers made for Charlotte Pass heavy winds blew over a two-storey brick wall while it was being built in Melbourne's Lygon Street.[28] Workers noticed 'a "creaking" sound, and felt the wall begin to sway'. Crashing down on an adjacent building, the wall caused considerable damage, especially to a bricklayer 'who fell with the wall through the roof of the shop and lay half-stunned, bleeding, and partly buried among the wreckage'. The front brought 'a heavy rain storm' to

Ballarat, where on wet roads a motorcyclist collided with 'a two ton motor van'.[29]

As the front swept over the Snowy Mountains, it whipped up a blizzard. The four skiers at Etheridge Ridge abandoned the idea of summitting Mount Kosciuszko, choosing instead to race back to safety at Betts Camp.

Fleeing this ferocious turn in the weather, Douglas, Larsen, Bottrell and Sodersteen struggled through a darkening landscape of snow and rock. They were learning firsthand why Aboriginal people had avoided Kosciuszko during winter. Even with skis, it is impossible to outrun a storm.

Navigating boulders and scree slopes and creeks, with snow building over marshes and crevices, any winter journey among the peaks was hard enough under clear skies. But in the blizzard's whiteout things got even harder. Edges lost their definition, angles became harder to judge. The visible world grew smaller.

Recounting his journey back, Sodersteen described how 'the wind began to rise and cloud came down on us'.[30] It got colder and darker and the going harder. At one point he slipped, tumbling downhill at Etheridge Ridge 'over and over for nearly 75 feet', only stopping 'when my skis jammed on a mount of ice', wrenching him to a halt. He was without serious injury, but admitted to being 'thoroughly shaken' by the experience.

The turn in the weather was so swift that the four companions were in imminent danger, as Sodersteen later explained:

> We began to hurry back and were soon in an icy blizzard. The
> sleet cut into our faces. We could not see any distance ahead,
> and urged each other to keep together, so that our progress

was slowed up, as two of the party were all out to it. With difficulty we reached Charlotte Pass hut. It was snowed up, so we dug our way in, and there we began to thaw out our hands, which were frozen and swollen. We were all famished. We had no food with us, so did not like the prospect of staying in the hut. We decided to venture out again and make for Betts.[31]

Having reached the tiny meteorological shed, they knew it would hardly suffice if they were trapped for a period of days. Betts Camp was a safer option, even if the journey to reach it was dangerous. Continuing onwards, the fleeing party struggled to stay together. Two of them started 'making down the wrong gully' at one point and were only saved from this mistake when one of the others went looking for them.

Finally, about three and a half hours after the blizzard first struck, they safely reached Betts Camp. Relief turned to worry as they entered. Evan and Laurie were not there.

The First Responders

After finding the hut empty, the men rested briefly. Then they went back out into the blizzard to look for signs of their missing friends. Bottrell and Larsen 'retraced their tracks' towards Charlotte Pass, probably hoping to meet their missing mates.[32] Heads bent to the ground, following their own ski tracks, they will have been aware that the snowfall was already obscuring the marks they had only recently made. And if they were out too long, they might easily get lost themselves.

Even expert skiers could only do so much in blizzard conditions. Bottrell and Larsen were unable to see very far, and the search became more dangerous the further they strayed from the hut.[33] Against the odds they managed to find some evidence of their missing companions. Laying in the snow, about halfway to the pass, they found 'a scarf and a glove' apparently discarded by one of their missing friends. Nothing obvious was nearby, which led them to conclude the clothing had been dropped much earlier in the day when the weather was fine. With 'the rapidly failing daylight and the severity of the weather', they turned back towards Betts Camp.

Meanwhile, Sodersteen and Douglas searched an adjacent valley:

> We stayed out until all was inky blackness, and kept coo-
> eeing, but the blizzard was terrible and we realised that if
> the missing men were not in the vicinity they would have no
> chance of picking up their right direction.[34]

They too were forced to return to the hut.

Getting Help from the Hotel

The blizzard passed with the night. Douglas got up early on Wednesday 15 August and set out alone to raise the alarm. Although the day was clear, a big dump of fresh snow made the going difficult. Forced to give up skiing after the first few miles, the ex-soldier 'discarded his skis and ploughed through on foot'.[35] Trudging through thick snow, he reached the hotel by 'strenuous effort' early enough that many of the hotel's guests had not yet left for their own skiing excursions. News that Evan and Laurie were missing spread quickly, and a search and rescue expedition was formed and dispatched.

Headed by a Norwegian-born ski champion named George Aalberg, this group headed towards Betts Camp. They aimed 'both to reinforce, and, if necessary, to relieve those who were already there'.[36]

Aalberg was known widely among the Kosciuszko crowd for being one of Australia's first ski instructors. Although he came to Australia to seek employment as a carpenter, first working in Canberra, he was soon enough professionally engaged by the Kosciusko Hotel.[37] When he was invited there as a guest of the Kosciusko Alpine Club in 1927 it proved a great leap forward. As one club member recounted:

> Everyone enjoyed his thrilling exhibitions of jumping, but of
> more use to the Club and its members was the good work
> he did in demonstrating the various swings, turns and stops.
> He also gave us many very useful hints on the selection and
> preparation of ski.[38]

In that same season he outdid even the most experienced local skiers by a huge margin, landing some '75 and 80 feet' from the jump point, twice the best length of the next-best skier.[39] And he won the championship by casually running at the back of the pack for the whole race until he 'was able to pass the field on the downhill run, winning the race' with apparent ease.[40]

In the weeks leading up to the rescue mission, Aalberg had been busily breaking skiing records and winning competitions. Summitting Kosciuszko from the hotel and returning in a single trip, he successfully shaved thirteen minutes from the previous record, completing the difficult 35-mile journey in under eight hours.[41] Aalberg had also won a recent cross-country championship, and broke the record for the run from Kosciusko Hotel to Betts Camp.[42] If anyone was well suited to lead a search party, it was him.

Before heading to Kosciuszko for the 1928 season, Aalberg had even attracted considerable attention in Sydney. Fiddling with a motorcycle he demonstrated how skis could be attached to them to make them operate in the snow.[43] He reportedly planned to attempt summitting Kosciuszko on this modified vehicle, but it was apparently not yet available for the rescue mission.

Aalberg headed straight to Betts Camp and found the hut empty. This was hardly a surprise, as the other members of the original party had gone out searching for their missing friends. But they had suffered some misfortune. Aalberg found a note in the camp that stated that 'Larsen is sick in Charlotte Pass hut'.[44]

After finding this note, Aalberg made for Charlotte Pass, but he soon met Larsen and one of the other men, either Bottrell or Sodersteen, making their way back towards Betts Camp.[45] They reported having spent the morning climbing Charlotte Pass, and 'ran down to the Snowy River, and turned downstream', effectively looping towards Perisher where they suspected Evan and Laurie might have gone. But when this proved fruitless they doubled back, retracing their own route back over Charlotte Pass towards Betts Camp.[46] Zigzagging to cover as much ground as possible, these exhausting preliminary searches found nothing.

If not before, then Aalberg's team now split up. There were various isolated stock huts and alpine shelters around the wider area, which were the most obvious places to search in the first instance. Everyone assumed that Evan and Laurie would have sought the nearest shelter, wherever they went.

Aalberg set off for what was known as the Tin Hut, situated 'near the crossing of the Snowy River at its junction with Spencer's Creek'. Two years previously this hut had saved a group of cross-country skiers who, like Evan and Laurie, faced a sudden change in the weather.[47] It was the kind of experience that gave the search parties hope, but also underlined the urgency of finding men known to be without food. In 1926 even with a well-prepared expedition, cold and hunger had proved hazardous, as one of the lucky survivors related:

> Visibility was very bad, and it became a grim fight, not only to reach the sheltering hut, but to find it. ... Hardly had the climbers settled themselves down in the Tin Hut when the blizzard burst in great fury. They had provided themselves with three days' rations, and in the hut were a tin of bully beef, a bag of flour and a tin of baking powder. ... Cramped for space, and with only one blanket to each man, and with the sodden wood throwing off more smoke than heat, the party suffered intensely from the cold. As the days sped by, and the raging blizzard prevented escape, their plight became critical.[48]

Eventually starved out, this group braved the persistently bad weather of 1926 and safely reached the hotel. As the search parties set out looking for Evan and Laurie, they were hopeful that the missing men would similarly turn up after sheltering somewhere during the blizzard.

But Aalberg did not find any sign of Evan or Laurie at Tin Hut. Nor did any of the other parties that searched other obvious shelters. As parties returned after a day of searching, the news from Betts Camp looked grim.

That evening, the story spread. A late afternoon press report detailed the news of the missing men, the first rescue party, the imminent departure of more rescue parties, and the notification of police.[49]

Rescue Reinforcements

The search and rescue operations grew as day broke on Thursday 16 August. George Aalberg and others continued their searching with the dawn. Some, including Aalberg, had reportedly spent the night in the isolated huts they were checking. Aalberg headed again for Charlotte Pass and then north towards the Blue Lake.[50] These early cross-country teams were trying to cover the various routes between key destinations, but their operations seem to have been dictated more by their own intuitions than any overarching plan. The consensus still seemed to be that Evan and Laurie had headed north from Charlotte Pass, aiming to loop back to Betts Camp by following the course of the Snowy River, then back up Spencers Creek. Such a route would take them around Guthrie Ridge, past the Perisher Range, and right back to the hut.

With Aalberg and others focused on the outlying huts and areas, the main body of would-be rescuers based at Betts Camp began coordinating with the Kosciusko Hotel. During the night two men left the hotel 'after tea … to grope a way in the pitch darkness to Betts' Camp'.[51] One of them arrived with a plan, clearly hatched at the hotel, 'to organise the searchers there – about 16 men – into parties, and to scour the numerous valleys and ravines'.[52]

With the morning, yet another party left the hotel for Betts Camp. This group had 'a strenuous time' making the journey. Bad weather was becoming a problem again. While pushing towards Betts Camp, this morning relief party from the hotel 'encountered a blizzard, with bitterly cold sleet driving into their faces'. Their vision was greatly restricted, and one of them 'was blown over twice'. As well as making the journey to the search area difficult, these blustery conditions meant that the search operation itself was about to become even harder.

Ideally, the search parties would have gone on the peaks and ridges. From there they could have looked down into the valleys for tracks or other signs of Evan and Laurie. But low clouds and dense precipitation made this tactic pointless.[53]

Unable to survey wide vistas from the heights, their best chance seemed to be an arithmetical one, playing with geometry and probability. They aimed simply to cover as much ground as feasible. But even this approach was becoming increasingly difficult. The immediate vicinity of Charlotte Pass had not yet provided any leads, so the searchers 'were convinced that they must be further out', but the further from Betts Camp the search area got, the bigger it grew. Evan and Laurie were somewhere in a growing circle measured in hundreds of square kilometres of snow and rocks and peaks and valleys.

At least the connection with the hotel was being maintained, and the welfare of the search parties was being considered. Fortunately for the still-unwell Larsen, this last relief party included Dr Archibald R. H. Duggan, the young and sports-keen superintendent of Sydney Hospital.[54] Larsen was suspected of having an 'internal haemorrhage', earned in the early searching for the missing men.

By now the rescue operation was becoming a much more organised affair. As the *Sydney Morning Herald* explained, this involved the rapid establishment of a considerable logistical apparatus:

The food supply for the searchers is a problem, as everything has to go forward by knapsack. The camp will only accommodate 16 men, so that as the search parties return to the camp more or less exhausted they are returning to the hotel, and their places filled by fresh men. Two men have also been detained to go backwards and forwards to the camp, taking food and stimulants, and bringing back the latest information.[55]

The newspaper also took its own initiative, ensuring its own supply of up-to-date information through the use of what it called 'our special reporter'. If they were not already there, this special reporter was most likely sent towards the Kosciusko Hotel either Wednesday night or Thursday morning. The Kosciuszko story was becoming a sensation.

News of Evan and Laurie

The *Sydney Morning Herald* broke the story on Thursday morning with the headline 'TWO MEN MISSING. LOST IN BLIZZARD.'[56] At least in print, they had got the scoop. Even as searching operations around Kosciuszko were continuing on Thursday, readers in New South Wales were learning the first substantial news about Evan and Laurie's misadventure.

Using reports from late Wednesday, the *Sydney Morning Herald* gave its Thursday readers a basic outline of the disappearance of Evan and Laurie, the preliminary searches, and the foreboding fact that 'Two members of the Millions Ski Club have been lost for 24 hours on the upper slopes of Mount Kosciusko.' For contemporary readers this was not just something that had happened. It was *happening*.

That immediacy gives the reportage a raw edge, but also makes it a complex source of historical information. As other newspapers jumped on the story, for instance, their rush to print could introduce error. The *Goulburn Evening Penny Post*, after hastening to get it into their Thursday edition, copied the story from the *Sydney Morning Herald* mostly verbatim.[57] They changed the headline to 'LOST IN SNOWSTORM Two Men Missing At Kosciusko', adjusted the date of the source information to Thursday, changed 'to-day' to 'yesterday', dropped the final paragraph about plans for Thursday, and made no mention of the *Sydney Morning Herald* at all. Such adroit plagiarism aside, it pointed to the way that information filtered out through wires and papers in repeating chains of information that could rapidly and widely disseminate a story, but also muddle its details. If journalism really is the first draft of history, then it is rarely clean copy.

In Sydney itself, where the afternoon and evening newspapers relied on the wires to add the most recent information to their stories, and where there was great competition to have the latest news, there are firmer timelines for information and better provenance. During the course of Thursday 16 August 1928, for instance, the *Sun* ran an article to which the editors appended short updates they got from the Kosciusko Hotel and the alpine town of Jindabyne:

HOTEL KOSCIUSKO, 1.45 p.m.
The Searchers are still out, and nothing has yet been heard of the result of their work to-day.
JINDABYNE, 3.15 p.m.
Some of the rescue parties had returned by about 3 p.m. to-day, but all told the same tale – 'Not a sign.'[58]

By the time this edition went to print, the editors had sourced a photograph of Charlotte Pass, which as the caption explained, was 'where the missing men were last seen'. The newspaper also carried a small hand-sketched map, helping readers to understand the geography of the unfolding events. A photograph of a bespectacled man standing on skis in the snow further illustrated the story. It was captioned 'Mr. Evan Hayes'.

As the day progressed without news of a successful rescue or so much as a sign of the missing men, interest in the story grew. Sydney's *Evening News* ran a large headline announcing 'TWO SYDNEY BUSINESS MEN LOST IN BLIZZARD'.[59] The subheadings capture the way the story was tapping into a sense of frantic searching and crisis: 'ALARM AT KOSCIUSKO ... SKI-ERS TRAPPED IN STORM ... SEARCHERS SCOUR SNOW WASTES'. It even carried two relevant photographs on its front page. One depicted 'DESOLATE COUNTRY WHERE MISSING MEN WENT', the other the hut at Betts Camp.[60] 'Late this afternoon, no word had been received at the Hotel Kosciusko regarding the missing men', it stated, above photographs of Evan and Laurie. The editors of the *Evening News* had now managed to source pictures of both men.

Further down this page was another photograph, of a small alpine hut. This, as the caption explained, was 'Pipers Creek hut, for which it is expected the missing men may make'. Even at this stage there was hope the men would be found alive. After all, the reports from the hotel could travel much faster and further than the latest news could reach the hotel from the rescue parties in the field.

The head of the Government Tourist Bureau shared an optimistic view that the men would soon be found or return safely of their own accord. Herbert John Lamble, who had done much to develop Kosciuszko as a skiing holiday destination, was also a Millions Ski Club Member and knew Evan and Laurie personally. He spoke to the press on Thursday, in time for some of the later papers:

> 'They might reach Piper's Creek or Digger's Creek, both of which lead up to the Hotel Kosciusko,' he said. 'The manager of the hotel has organised a committee of ski-runners, which includes experienced bushmen, and efforts by individuals have been forbidden. They will explore right up the Snowy River from Jindabyne. There is another exploring party out from Kiandra, and a third along the Crackenback Range.[']61

Photographs of alpine search parties, printed side-by-side in Thursday's edition of the *Sun*, gave readers some visual impressions of these rescue operations.62

As the story continued to unfold, the assistant director of the Government Tourist Bureau, Lamble's colleague John G. Cocks, attested to the missing men's skiing abilities:

> 'I know Evan Hayes personally,' said Mr. Cocks. 'For three or four years now he has been to Kosciusko, and he and Seaman are both regarded as expert mountaineers. Both have made the trip to the summit of Mt. Kosciusko from Bett's Camp several times before.'63

While slightly misinformed about Laurie's Australian skiing history, Cocks was stressing that the men were as well-placed as any to endure trouble in the Kosciuszko area. But he also underlined the difficulties of such mountain skiing:

114

> It takes a good deal of nerve and physical endurance to
> undertake this trip. The arduous climbing in herringbone
> fashion – one ski across and in front of the other – up in the
> snow-covered mountain, would tire any but the strongest and
> much skill is required in the return journey in running down
> the slopes. Some of the runs are about a mile long, and the
> ski-runner travels down at a speed of anything from 25 to 35
> miles an hour.[64]

The public servant was partly underlining Evan and Laurie's
strength and skills, but probably also explaining that the searching
itself was difficult. He was evidently trying to manage the story:
a tragedy at Kosciuszko was not good news for the Government
Tourist Bureau.

Understandably, the Sydney press showed the most interest in
the story. Evan and Laurie were residents of the city, and Kosciuszko
was one of the geographical gems of New South Wales. But, typical
of the period, the story travelled swiftly along the wires that
networked the nation and the world. Melbourne's *Herald*, *Argus*
and *Age* soon carried it, as did Adelaide's *Register* and *Advertiser*.[65]
Other capital city papers presented the story to readers in Brisbane,
Hobart, Canberra, Darwin and Perth.[66] Regional papers also piled
in, giving the latest reports to readers of Wagga Wagga's *Daily
Advertiser*, Mount Gambier's *Border Watch* and Townsville's *Daily
Bulletin*, as well as many others.[67] As their disappearance was
widely reported, and the rescue efforts followed by a fascinated
public, Evan and Laurie achieved national fame.

The story also broke national borders. As early as 16 August it
was reported in the *Auckland Star* and Wellington's *Evening Post*.[68]

As with Australia, other New Zealand papers soon followed. In this world stitched together through a growing electrical network, the news front can be observed much like the weather, showing the story of a blizzard at Kosciuszko fanning out through the world. Eventually, it reached Laurie's home state of New York.[69]

Amid so much reportage, and across such wide coverage, the facts on the ground about the missing men remained slim. Even now, the events of those first few days are hard to untangle. While the Kosciuszko tragedy produced an abundance of reportage, feeding public fascination, the facts got buried by much irrelevant commentary.

But beyond the story of an incident, and past the theories and speculations, the wider coverage reveals something special. A cast of characters were illuminated by this moment of crisis. As the search for Evan and Laurie continued, through their would-be rescuers we capture a momentary glimpse into a young nation's soul.

CHAPTER FIVE

Search and Rescue

Australia's alps demand people work together or risk death. This was so from Kosciuszko's earliest documented moments when the bushcraft of Charlie Tarra kept Strzelecki from disaster. Undoubtedly it was part of Kosciuszko tradition long before that. Even in legend it rings true. Above all his other attributes, Paterson's man from Snowy River most personifies charity. Riding out from Kosciuszko, he came to aid others.

The striking landscape can make a person feel small, insignificant even. But with its strange loneliness, there also comes a certain type of community-mindedness. Even now walkers and skiers urge each other on with a good cheer, strangers helping strangers. When Evan and Laurie went missing, the unfolding tragedy helped foster a momentary sense of national consciousness.

Back in 1909 a reporter who covered the opening of the Kosciusko Hotel recognised something of this phenomenon, starting his report by conjuring 'the last stretch of 50 miles of slushy and mostly snow-covered road' with snippets of overheard conversation within the coach.[1] The quoted banter was in clear contrast to the destination, which he described as

> just one great lonely hotel, a big mass of a building dumped
> all by itself on a snow-covered hillside, some degrees more
> isolated than the woolshed of a Darling River run. ... built
> high up on the side of Mount Kosciusko far away up the
> ragged, erratic valleys on part of the wild original world
> where no man would dream of going were it not the very best
> place to leave the world and its worries ...[2]

With these words he captured that ironic effect: a landscape heightening a sense of isolation while throwing people into close confines. Although he did not use the word, he might have thought it was a landscape peculiarly adapted to fostering that thing Australians tend to call 'mateship'. Especially so, perhaps, considering this journalist was Charles E. W. Bean, later Australia's official war historian.[3]

That Kosciuszko had a special role in fostering a sense of national community became especially evident during the First World War, where it served as an iconic shorthand. Nationalistic poetry, for instance, tapped into the popular sense of community engendered by the place and its mythic embodiment of Australian ideas. 'The Men From Snowy River', a poem clearly playing on Paterson's ballad, began with this collective unity in mind:

> They're marching from the mountains – from Snowy River
>   side,
> Heard the call of Empire one and all;
> Kosciusko sent the message far and wide,
> And the men from Snowy River are responding to the call.[4]

Once abroad, soldiers and journalists understandably used Australia's most famous mountain to help explain what they saw or felt in their letters home. One account of the landing of 25 April 1915, for instance, pointed out that Gallipoli's 'ridges resembled the slopes of the lower portions of Kosciusko'.[5]

As an icon of Australia, Kosciuszko also became a talking point with allies, illustrated by one soldier who wrote about how 'English folk of education and standing were amazed at pictures of Illawarra and the Hunter Valley and Kosciuszko'.[6] The mountain was clearly a talking point to upset stereotypes. The same writer continued by pointing out that 'the majority of Englishmen believe we have no song-birds and no bush flowers and no snow'.[7] That he was carrying pictures of Kosciuszko further highlights its significance as a memento of home.

Kosciuszko's emotional pull was evidently great. Another soldier referred to a mountain near his position as 'a Kosciusko of these regions' and admitted the very sight of it 'intensifies a nostalgia for the Bogongs'.[8] His musings were prompted by receiving a pair of socks from an Australian comfort fund in 1917, and in fact wartime fundraising events sometimes played on the cosy community for which Kosciuszko holidays were renowned. A 'Kosciusko Carnival' in Lismore in 1918 created a snowstorm effect with electric fans and confetti while collecting money to provide beds at the local hospital for wounded soldiers.[9]

A popular wartime song composed by Ella Airlie, 'Back to Kosciusko', especially captures this sense that the mountain was an emblem of Australia. 'A batch of wounded soldiers sent to England', it started, talked longingly of home:

Far away, where icy winds blow, lives a little maid that I
know. She whom I idolize and every evening when the shades
are falling I can hear her sweet voice calling come back my
love to me. … Oh! I must go to Kosciusko. Back to the land of
snow. For my love's in Kosciusko waiting me I know and her
brown eyes, with lovelight beaming, will soon with rapture
glow. Then wedding bells will start a ringing telling of the joy
they're bringing, when I'm in Kosciusko. Little mountain maid
I'm lonely, for I love and want you only. My bride you soon
will be. Beside the mountain we will dwell for ever. And I'll
never leave you, never in other land to roam. …Oh! I must go
to Kosciusko….[10]

This wartime longing for Kosciuszko survived the war's end as a
powerful motif. In 'Anzac Day 1926' one poet asked a dead soldier:

Are you dreaming of Australia on this still sad Anzac night,
Of her deep ravines and gloomy gorges wide,
Of the snow-capped Kosciusko gleaming 'neath a smiling sun,
When winter frosts the range of Great Divide?[11]

Yet for all its unifying effect, Kosciuszko also uncovers community
divisions. Again the war proves instructive, as when a German
was sacked from the Kosciusko Hotel because of political agitation
about the employment of an 'enemy alien' in 1915, or when returned
servicemen in Cooma objected to the foundation stone for a new
building being laid at the hotel on Anzac Day in 1926, which they
felt detracted from the unveiling of Cooma's war memorial.[12]

Most unifying of all, however, were the search and rescue
efforts of Kosciuszko. These showed how harsh the Kosciuszko

landscape could be. When a group of returned soldiers summitted the mountain on skis in the winter of 1919, their achievement was newsworthy. As one report put it, 'the diggers were at the top of their physical powers; they had been training (at another sport) for four or five years in Gallipoli and France', and were in peak physical condition.[13] Yet some had to abandon the attempt on the mountain through injury, others were reduced to blindness, one nearly got lost, and another rather gracelessly slid down the mountain on his bottom as the safest means of retreat. If the mountain could humble mighty Anzacs, then Evan and Laurie's would-be rescuers needed to be tough.

Police en Route

An impromptu committee of men at the Kosciusko Hotel had directed the early search and rescue operations for Evan and Laurie, but when they had no success, they had notified the police. The small village of Jindabyne, which had the nearest police station, received the first news of Evan and Laurie's disappearance by wire from the Kosciusko Hotel at about 7.30 pm on Wednesday evening, well over thirty hours since they had last been seen.

By one account, a preliminary search party led by Mounted Constable Alfred James Lambert set out at once. Lambert's team 'proceeded to Thredbo Falls to the head of Friday's Flat', checking one of the potential routes that Evan and Laurie might have used to get down from the highest ranges.[14] Meanwhile, Sergeant James E. D. Carroll, the ranking officer at the station, arranged for a more thorough search to begin on Thursday morning.

The Jindabyne police had a lot of experience conducting mountain rescues. They were used to being called upon to help with various alpine crises. Whether by horseback or on foot, their rescue expeditions had changed little from the early days, save for being able to benefit from the construction of the Kosciusko Hotel, which meant that they could wire for help and take to the field faster.

One of the Jindabyne police's best-recorded missions from the 1920s occurred in June 1926 when they went to the rescue of a herd of cattle 'snowed-in on Mount Kosciusko'.[15] Fortunately for the cows involved, a prominent member of the Royal Society for the Protection of Animals happened to be holidaying at the Kosciusko Hotel. When word arrived of the animals' plight, he sprang into action, organising funds for a rescue operation, which the Jindabyne police then mercurially agreed to undertake. They set out from the hotel that afternoon:

> Night was closing in, and a blizzard was sweeping across the hills. Then men went single file on a nine-mile tramp. Their snow-shoes were slung across their backs, as the men could not use them. As they tramped on the snow deepened, and after going five miles the wind blew fiercely. The men were all wet through, and could not see one another. Occasionally they would bog up to the hips in drift snow, or fall in the snowed-over creeks, the water in which was extremely cold. Ultimately, they reached Bett's Hut, about 10.30 p.m, all done in, … A 'billy' of tea was made, and the men lay round the fire until they were aroused at 4 a.m.[16]

Continuing with the morning, this cattle rescue-mission found their first animals 'frozen-in at the side of the river, and covered with snow'. Only fox-tracks had led the party to the spot where the corpses were buried under snow.[17]

The party continued, managed to find sixty-three cattle that were still alive, and herded them back to safety. It was a win for the Jindabyne police. A little over two years later, however, the stakes were much higher.

On Thursday morning 16 August 1928, three days after Evan and Laurie left the hotel, Sergeant Carroll led a rescue mission. This attracted less media attention than the cows of yesteryear, largely because it proved unsuccessful. Yet, remarkably, it travelled as far as the Victorian border.[18] According to information gleaned by the *Sydney Morning Herald*, Carroll's party

> followed the Thredbo River on the Victorian side, down as
> far as Grogan's Gap, over which Mount Kosciusko towers.
> They were on horseback, and said that frequently the horses
> were down to their girths in snow. At other times, they were
> up to their knees in boggy marshes. They had travelled over
> 70 miles, penetrating mountains, and with great difficulty
> they made for the spot [where they spotted smoke], only to
> find it deserted. Evidently some shepherd had been travelling
> overland.[19]

And so the search continued.

Further Reinforcements

Other outlying police stations were also asked for help looking for Evan and Laurie, including from the small village of Kiandra, 75 miles from Kosciuszko.[20] Formerly a significant gold mining town, Kiandra's heyday had been the early 1860s. By the late 1920s the miners had mostly gone and the settlement's surviving buildings were described by one writer as 'weather-worn and decayed'.[21] But although small, in terms of winter mountain rescues, Kiandra could more than hold its own. It was, after all, the place where Australian skiing reportedly began.

By one account a Norwegian miner 'brought the secret of the Ski' to Australia with his arrival in Kiandra in 1886, introducing more modern ski manufacture.[22] But simple skis were reportedly first seen around the Kiandra area as early as 1857, supposedly introduced by a Swedish-born resident.[23] Either way, a domestic skiing culture thrived there, such that Kiandra residents sported a distinct type of ski 'flat from heel to bend', and exhibited their own distinct Kiandra style of skiing. Having adopted this mode of transportation for its utility, Kiandra folk justified their use of simple skis for very practical reasons:

> They argued that a tightly-attached ski in deserted country
> might mean a broken ankle and death from exposure. So
> they simply kicked the foot into a toe-thong and went off to
> round up the cows. Business, not pleasure, was their need,
> and probably they were right. In their isolation safety was
> indispensable. Even now [1927] the Kiandra postman does his
> 15 miles a day in winter on skis.[24]

Beyond utility, the Kiandra skiers were also reportedly good sportsmen, being particularly famous for ski-jumping. When news reached Kiandra about missing men at Kosciuszko, such locals were the sort who heeded the call when their local police formed search parties.

The Kiandra locals also had wider regional experience. Kiandra had served as one of the main tourist entry-points to the snowfields during the early twentieth century before it was eclipsed by the establishment of the Kosciusko Hotel. Even afterwards, it remained a significant point for skiing enthusiasts. From the mid-1920s ski parties were making occasional winter overland journeys between Kiandra and Kosciuszko, prompting calls for more shelter huts.[25] As with rescue teams operating closer to the Kosciusko Hotel and Jindabyne, searchers from Kiandra probably headed to isolated huts first. They presumably also sent whatever resources they could spare towards the Kosciusko Hotel. Little evidence remains of the efforts of these early police parties from Kiandra itself.

Much the same is true of the Victorian stations that were also called upon to help find Evan and Laurie.[26] Police stations near the border between New South Wales and Victoria would have swung into action swiftly after being contacted by the Kosciusko Hotel directly. There were even calls from Melbourne to contact and divert a Victorian cross-country skiing expedition, which was already in the Victorian alps when the news broke.[27] This team was travelling towards Mount Bogong, Victoria's highest mountain, and it was thought that it could be diverted towards Kosciuszko to contribute its members' expertise to the search. The president of the Sydney Millions Club, Sir Arthur Rickard, happened to be visiting Melbourne and reportedly began assessing the feasibility of this option once he learned the news.

In Sydney, where the missing men were best known, and where the Cooma Mail train provided ready mountain access, other rescue parties were also forming. When word hit the city that Evan and Laurie were missing, several men immediately volunteered to go up to the mountains and join the search. They organised swiftly and departed quickly. A group of them was photographed at Central Station prior to boarding their train, displaying large packs and long skis.[28] 'There was an animated scene at Sydney railway station ... as the Cooma mail train left', the *Sydney Morning Herald* reported the next day.[29] Reminiscent of the children's narratives of travel, the departure from the station was a moment of considerable fanfare, especially because the group was composed of expert skiers. By witnessing their departure and wishing them well, Sydneysiders may have hoped to participate in some small way, to feel closer to the story.

Teamwork and Tea Ladies

When the Cooma Mail train left Sydney, there was no sign of Evan or Laurie. There was still no news when the fresh rescuers reached the Kosciusko Hotel about 9.30 am on Friday 17 August. An hour later, one of the rescuers sent a telegraph to his father. It read: 'Leaving for Betts' Camp immediately. Will be away two days.'[30]

Soon thereafter, this team headed towards Betts Camp 'to take the place of the other men who had returned', arriving 'about midday'.[31] Fortunately, the route from Kosciusko Hotel to Betts Camp was becoming easier now, aided by the provision of a refreshing station. As one report mentioned, 'Miss Barber (Victoria) and Miss Gelling, who are excellent ski-ers, have done good work in establishing themselves half-way to Bett's Camp, and are supplying hot tea and sandwiches to the workers as they pass backwards and forwards.'[32] Although these women were only mentioned in passing, there is no reason to believe that women were not more actively involved in less-documented search parties.

In fact, the snowfields were increasingly crawling with searchers, as holiday-makers joined the wider rescue mission. As a report wired from Jindabyne that afternoon stated:

> There are very few men left in Jindabyne to-day. Most of the
> able-bodied men in and around the town to the number of 40
> or so have joined in the search under Sergeant Carroll.[33]

By late Friday it was also reported that 'over 100' individuals
were engaged in searches from the Kosciusko Hotel.[34] This sort of
regional mobilisation had probably not been seen since the war.

During the afternoon the party of reinforcements from Sydney
re-crossed some of the ground that had already been searched. They
focused on 'the whole of the western side of the Perisher Range,
including the Snowy gorge', basically still covering the area through
which they expected Evan and Laurie would have to pass if trying
to reach Betts Camp from a northern direction. The Jindabyne
police also reached Kosciusko Hotel on Friday morning, reportedly
arriving directly by car, before also heading into the field. They set
off from the hotel 'at midday to cross the Perisher Range down to the
Snowy', furthering the focus on this same general area, but coming
at it from the eastern side nearer the hotel.[35] In essence, the searchers
were focused on the main ranges immediately north of the road.

But this was still a huge search area, and each search party had
to undertake considerable travel even before their searches could
begin. Like an army with over-extended supply lines, the rescue
force was severely limited in its capacity to push deep into the search
area for longer periods. That a police party left the hotel at noon
and was not expected back at the hotel until roughly midnight,
making for twelve hours in the field, neatly captures the scale of
search operations.

Even with a logistical apparatus making it easier to stay in the
field, things were looking grim. Three nights had passed since the

blizzard. The search area was enormous. There were no new leads coming from the search parties, still focused on the ominously named area around Perisher. Needing more help, the rescuers looked for it from above.

Aerial Search and Rescue

At about 10.30 am on Friday 17 August 1928 a Moth aeroplane took off from Mascot in Sydney and headed south towards Kosciuszko. Such alpine flying was still relatively rare.[36] It was only several years since a plane had first flown over Kosciuszko, and only a few weeks since one of the most famous such flights.[37]

The Moth from Sydney was piloted by Rupert King. He had successfully demonstrated the feasibility of aerial searching earlier in the year.[38] On that occasion the 'famous architect and engineer of Sydney', Henry E. White, had been missing for three nights. Having taken his yacht to sea, the missing architect intended heading for Port Stephens, but failed to arrive when expected. Although he was an accomplished seaman who had sailed the yacht from New Zealand to Australia, there were concerns for White's safety, so King was dispatched to investigate. Within two hours he found the missing yacht, which was anchored safely not far from Port Stephens. All aboard gave 'reassuring signals', and two hours later King landed back in Sydney with the good news. Although hardly spectacular, aerial search operations had thus already proven feasible. There were high hopes as he took off for Kosciuszko.

The alpine conditions were demanding, but King was suited to the mission. A highly skilled pilot, he was also instrumental in developing early Australian aviation.[39] Fascinated from childhood with flying, he tried to enlist in the Australian Flying Corps in

1915 but was rejected because of poor eyesight. Undeterred, King travelled to England at his own expense and successfully joined the Royal Flying Corps. For the first few years of service he was kept grounded as a driver, but pestered his way into flight training in 1917, where his learning-curve was steep:

> We had been flying around and then, when we landed, my
> instructor hopped out of the aircraft and took his crash
> helmet off. I asked what he was doing. 'You're going off on
> your own now' he said. I said 'Don't be bloody silly! I've only
> done three and a half hours.' 'Yes, you're alright – off you go!'
> And off I went.[40]

King was eventually posted to Egypt and Palestine and joined bombing raids against Turkish positions. After the war he returned to Australia and became one of the early members of the new Australian Aero Club, formed in 1919. He was employed as a flying instructor for the club in 1926. A few years later King was a co-founder of a new company, Air Travel Limited, which imported 'an American all-metal biplane, the first of its kind in the Commonwealth'.[41]

In the Kosciuszko-bound plane with King was another war veteran, the flight mechanic Walter Shiers, who had served with the Australian Flying Corps.[42] Shiers had eventually ended up in England, returning to Australia by what was a very unusual means, one that qualified him too as being well suited to the Kosciuszko search operation, especially seeing as he had experience of a wide range of alpine conditions.[43] As his war service record notes, he 'Left by Aeroplane for Australia 12.11.19' and 'Returned to Australia per Aeroplane arrived 10/12/19.' These annotations hide a grand adventure. With war's end, the Australian government had announced a competition for the first successful flight from England to Australia. The prize was £10,000, and Shiers departed England as a flight mechanic on one of the competing planes. Travelling with the pilot brothers Ross and Keith Smith and another mechanic, James Bennett, Shiers was one of the winning team, sharing a quarter of the prize money.

The Moth approached Kosciuszko in 1928 with great preparation and some caution. King and Shiers fitted it with additional fuel tanks, and the Shell company provided fuel depots ahead of their arrival. The nearest functional landing situation to Kosciuszko, at least for King's Moth, was at Cooma, so they headed there first. As the local paper described it:

> One aeroplane from Sydney left there at 10.30 this morning, and after passing through two storms landed at the Cooma racecourse at 2.30 this afternoon and proceeded to the search shortly afterwards.[44]

Later, King explained his search method with a reserve that belies the flying skill and technical competence required:

> I took Wally Shiers with me in the Moth and we went looking for them [Evan and Laurie]; flying up and down, just a few feet above the snow. I used Cooma as a base, landing on a piece of ground near the hotel where we stayed the night.[45]

There was, of course, no real airstrip.

Reporting on the adventure of the flight, the *Sydney Morning Herald* referred to some of the difficulties King and Shiers faced above the mountain:

The flight over Mount Kosciusko lasted almost two hours. Although a careful search was made from a height of about 700 feet, no traces of the men were found. Weather conditions were unfavourable, and the Moth had to face a severe gale, blowing over the mountains at about 60 miles an hour.

Mr. King returned to Cooma at 5.30 p.m.[46]

By that time another aeroplane from Sydney had also reached the mountains to search for Evan and Laurie.[47]

Some Scandinavian Success

While additional search parties arrived from Sydney on Friday, and searchers spotted the aeroplanes flying over the mountains, George Aalberg headed to Mount Kosciuszko itself.[48] Setting out from Betts Camp, Aalberg was accompanied by another accomplished skier, Swedish former soldier Ivor T. Soderlund, who, as a reporter noted, 'controls a big importing business in Sydney, and only recently joined the Millions Club'.[49]

After returning to the Kosciusko Hotel, Soderlund recounted their search to a waiting journalist:

Aalberg and I left Betts at 9 a.m. to-day under wretched
snow conditions, the snow being soft and sticky. We had to
push ourselves downhill. We were determined to try out an
alternative track to the summit, which Hayes and Seaman
may have taken. We crossed to the left-hand side of Charlotte
Pass, and did a tremendous amount of climbing work towards
the Snowy River. We came up to the right-hand side of the
summit, but saw nothing there, so we decided to turn down
to the left towards Merritt's Creek, as the wind was blowing

in this direction to-day, just as the blizzard was blowing in on the day the men were lost. We assumed that, unable to face the gale, the men had turned their backs to it, and tried to make down to a lower level.

About a mile from the summit we were ski-ing down into a valley, when we came within an ace of losing our lives. We were ski-ing down a long slope, when a drop of nearly 400 feet opened up in front of us. I just had time to fall and dig in my alpenstocks. Seeing me do this Aalberg followed suit. Otherwise it would have been the end of us.[50]

The near-miss was a salutary reminder of the dangers on the mountain.

But such close searching had its advantages, as Soderlund soon discovered:

Working round into this valley about two and a half miles from the summit we suddenly crossed big ski marks in a sheltered position. For about 50 yards the lines were frozen into the snow, and after that the drifted snow had covered them up.[51]

This stretch of exposed ski trails was the most significant find since the discovery of a scarf and glove on the night of the blizzard. To Aalberg and Soderlund's experienced eyes they looked to be 'two or three days old'. Moreover, there appeared to be two sets of marks.

Obviously excited, the Scandinavian pair 'kept on zig-zagging over the area' for about an hour and a half, trying to pick up more of the trail. But finding no further ski marks they had to turn for the Kosciusko Hotel in the early afternoon. They were a long way out from the other teams and needed to get the news back as soon as possible.

The exposed ski tracks were only partial, but they were sufficient to give a general directional impression, providing a vector of movement that the search parties could use to narrow the search area. It was a game-changing find. 'The tracks were going in the direction of the Crackenback range,' Soderlund explained to a reporter for the *Sydney Morning Herald*, who elaborated what this meant:

> The marks indicated that the skiers were trying to work
> down from the high ranges into the timbered country which
> borders the Thredbo River, some thousands of feet below.
> There are settlers in this district, the nearest being about
> 14 miles from Mount Kosciusko. Attention will now be
> concentrated on this locality, and parties are being organised
> to try out the surrounding country right down to the plains
> below. There are a number of huts there, which are used
> by shepherds in summer time, and it is hoped now that
> the missing men, knowing this, have made down off the
> mountain.[52]

When immediate enquiries around the hotel found no other skiers had been in this area recently, it was generally concluded that these were probably the tracks of Evan and Laurie. Finally having something substantial to work, Sergeant Carroll planned to send horsemen into the timbered country for which the tracks seemed headed. He also arranged to divert his most expert searcher, intending that 'a black tracker will pick up the ski marks and follow them if possible'.[53]

Tracker Rutherford

William Rutherford had been with the search parties from Jindabyne since at least Thursday, although he had received scant mention in the press.[54] One early story referred to him as a 'black tracker from Dalgety', a town a little south-east of Jindabyne where Rutherford was stationed.[55] Yet Rutherford was a well-known local personality who had worked with the police for quite some time, and his Kosciuszko roots went deeper than them all.[56]

To an extent Rutherford was part of that high-country horse-riding culture that seems straight out of a poem, and perfectly suited to a Kosciuszko rescue story. One writer, styled 'Aussie', described Rutherford as 'a great horseman', perhaps the greatest accolade in that time and place that could be given.[57] Moreover:

> He has often been the means of saving people who have been lost in the rugged country which he knows so well. To meet such men quickly banishes all unfavourable ideas of the Australian aborigines, and one wonders why more of his class have not been given equal chances.[58]

Another writer characterised Rutherford as 'a prominent figure, respected and liked by the law abiding and hated by the lawless'.[59] 'He knows the Snowy River country as few men know it', this writer added, also noting that 'his deeds as an officer of the law on the side of society are numberless.' Yet Rutherford obviously suffered from racial prejudice. This same writer pointed out that he 'was black, and his skin stood as a bar to rank as an ordinary member of the police force'. It was clearly this which relegated one of the great saviours of the Kosciuszko region to the position of nameless black-tracker.

To some extent the Kosciuszko rescue mission lets us see a man like Rutherford and the prejudices of his time and place, at least for a few moments. A former student of the public school at Delegate, his hometown further south-east near the Victorian border, Rutherford reportedly 'writes a faultless copperplate, and had knowledge of literature that few ordinary men possess'.[60] Yet while his own written words remain few, Rutherford likely experienced at least some of the same sort of discrimination found in the region by a traveller visiting in Rutherford's youth:

> Returning [from the recreation ground], we met Ned, the aboriginal, once a tracker, going forth with a gun and rod, accompanied by a little dusky son. 'Why is he not at school?' The boy looks shyly down. Later we hear that one of his race has been refused admittance to school in an adjacent hamlet on the score of colour. And so young Ned, fearful of similar treatment, instead of wrestling with the multiplication table, goes forth to shoot and fish with his father.[61]

Fortunately for Rutherford, he persisted with his education, clearly aided by a remarkable father.

Like Evan and Laurie, Rutherford was named William for his father. Like Evan, Rutherford's heritage can be traced back into Kosciuszko's past. But more than either of them, Rutherford's family story captured the complexities of the colonial experience. His father was taken by settlers near Berridale, trained to ride, played sports and was considered 'a buck-jump rider equal to the men from Snowy River with whom he was brought up'.[62] But when his foster-parent died childless the family estate went to a nephew rather than their adopted Aboriginal son.[63] Despite helping discover 'a gold-bearing reef' in 1900, the Rutherfords never escaped their subordinate social standing.[64] Although his father was a leader within the local Aboriginal community, Rutherford's family struggled, highlighted by a reply from the Board for Protection of Aborigines responding to one of William senior's letters:

> Sir, – With reference to your letter of 26th ultimo [September 1904], covering a communication from William Rutherford, an Aboriginal resident at Delegate, I am directed by the above Board to inform you that they are unable to authorise the issue of meat rations to the Aborigines at Delegate and it is reported that they can get as much soap as they required from surrounding residents. Rutherford himself is said to be a powerful man, about 45 years of age, who is well able to provide for himself and his family. With regard to the request for clothing it is represented that all the adults and children are fairly well clothed at present.[65]

For the Aboriginal communities of Australia, these were often difficult times, and legal strictures disproportionately affected their lives. The passing of the 1909 *Aborigines Protection Act* for New South Wales a few years after this correspondence created a wider range of rules and regulations governing Aboriginal lives.[66] Rutherford was in young adulthood by the time the *Aborigines Protection Act*'s provisions came into full effect with custodial guardianships, control of reserves and the like. But he had most likely witnessed the limits of the law for Aboriginal people even before this period. His father was called to court in 1901 to give evidence against a hotelier that had supplied him, 'William Rutherford, an aboriginal, with liquor – to wit, whisky'.[67] Serving

Lake Cootapatamba and Mount Kosciuszko (top right, behind ridge),
photograph by Nick Brodie, 2018.

At 2228 metres above sea level, Mount Kosciuszko (above) is Australia's highest mountain. A place of extremes (below), it lends its name to a wider landscape. But Kosciuszko is also an important part of Australia's history, culture and identity.

'Gathering storm over Kosciusko from Guthrie', photograph by Frank Hurley.

'The Ladies Ski Race, Kosciusko Carnival', c. 1900–1925.
STATE LIBRARY VICTORIA, ACCESSION NO: H82.43/44.

Domestic Australian skiing dated from at least the gold rush years. Kiandra (above) was one of its main centres. In the early twentieth century, organised ski tours to the Kosciusko Hotel became common. The Millions Ski Club (below) was one of many groups to travel there by train from Sydney via Cooma.

Millions Ski Club at Cooma Railway Station, en route to Kosciuszko, c. 1927. Evan Hayes appears to be seated front and centre. AUSTRALIAN SKI YEAR BOOK, 1928, P. 11.

'Kosciusko Hotel', photograph by Charles H. Kerry.
NATIONAL MUSEUM OF AUSTRALIA.

The Kosciusko Hotel (above, in winter) was the main ski resort in New South Wales during the 1920s. Betts Camp (below, in summer), was the closest accommodation to Mount Kosciuszko at this time, and the base from which most winter summit attempts were made.

Betts Camp, c. 1910.
NATIONAL LIBRARY OF AUSTRALIA, NLA.OBJ-147474125.

Evan Hayes.
Australian Ski Year Book, 1929, p. 156.

Evan Hayes was a competitive skier and is shown here with a selection of trophies. Laurie Seaman travelled the world as a photo-journalist. Chrissie Seaman was a wartime volunteer and active in charitable endeavours.

Laurie Seaman.
Australian Ski Year Book, 1929, p. 158.

Chrissie Seaman.
Sydney Mail, 12 April 1933, p. 40.
National Library of Australia.

Laurie Seaman standing beside a car. Chrissie is one of the passengers.
NORTHERN BEACHES LIBRARY SERVICE COLLECTION, COURTESY OF LAURIE SEAMAN.

In summer, Kosciuszko was accessible by car. Laurie (above) was a noted motoring enthusiast. In winter, dog sleds (below) were a popular attraction at the Kosciusko Hotel. The dogs were veterans of various Antarctic expeditions.

Antarctic dogs pulling a sled on Kosciuszko, c. 1900–1925.
STATE LIBRARY VICTORIA, ACCESSION NO: H82.43/48.

Dressed for an expedition, these acquaintances of Evan and Laurie were prominent in early skiing circles. Herbert Schlink (left) and Eric Fisher (right) both took part in search operations.

Doctor Eric Fisher and Doctor Herbert Schlink on skis, Kosciuszko, New South Wales, July 1927.

'A skier at Charlotte Pass during a blizzard, 1926', photograph by Herbert H. Fishwick.
NATIONAL LIBRARY OF AUSTRALIA, NLA.OBJ-163327307.

Kosciuszko's climate proved hostile to rescue parties. Those on skis faced their own blizzard conditions, similar to that shown above, while aerial searchers endured powerful winds. The image below was taken by one of the search planes a few days after Evan and Laurie went missing.

Aerial view over Mount Kosciuszko taken from the Gipsy Moth of Rupert King and Wal Shiers while searching for 'lost hikers', 1928.
NATIONAL LIBRARY OF AUSTRALIA, NLA.OBJ-144781653/VIEW.

Laurie Seaman (left) and Evan Hayes (right), photographed by each other at Mount Kosciuszko's summit cairn, 14 August 1928. Seaman's Hut, erected nearby, remains a poignant memorial to their subsequent misadventure.

Seaman's Hut, photograph by Nick Brodie, 2018.

alcohol to Aboriginal people had been prohibited in New South Wales since 1867.[68]

This context all helps explain Rutherford's joining the police, despite their seeming role in being agents of repressive law. Relatively poor, he needed the work. Aboriginal, his options were limited. Competent with horses, he was suited to the role. Tracking work helped him to stay connected to his country.

Hopes were high that Rutherford could follow the trails of Evan and Laurie because earlier that year he had acquitted himself well in a story that gained widespread notice. He had been called to help recapture two escapees from the Mila Afforestation Camp in March 1928. While innocuously named, the Mila Afforestation Camp was in fact a prison labour camp, a little south of Bombala, where inmates were put to work clearing forests and establishing pine plantations.[69] Relatively self-contained and isolated, it had echoes of the convict stations of the past.[70]

The prisoners escaped by cutting away part of a door frame. They also cut the telephone wire that connected the camp with the Mila Post Office as they fled into the surrounding forest.[71] But the town of Mila itself was not far from the camp, and the wires connecting the post office and Bombala township were still intact. Soon enough the Bombala police were on the hunt with, as reports characterised him, 'black tracker Billy'. Within a few days Rutherford found signs of the escapees beside a railway line. He led the police along a series of camping sites, gradually closing in on the escapees:

The tracks of two men were followed and, leaving the line, within half a mile, they found where two persons had camped. The tracks were again followed along the line to Jincumbilly siding, where they again left the line.[72]

A detachment of police was sent ahead, while Rutherford boarded a train with the lead sergeant and a constable. Soon enough the driver spotted 'two men sitting alongside the line' and halted the train. As *The Bombala Times* then detailed,

> The police jumped out and arrested them. They made no resistance, having evidently had enough. After their long tramp, mostly in wet weather, they were tired, footsore, and hungry. They were still wearing some of the prison clothes when captured.[73]

The escapees were sent to Goulburn Gaol, given an additional six months to their sentences, and Rutherford slipped back out of the pages of the newspapers until the next adventure, several months later in Kosciuszko.[74] There, his expertise faced a much more difficult challenge.

The Ski Marks

'Black-trackers and police parties, and bands of skilled ski-runners, have covered an area of about 100 square miles since the search began on Wednesday', wrote the *Sunday Times*' reporter late on Saturday 18 August 1928.[75] The operation had continued to grow, as even more volunteers poured in from Sydney by Friday night's Cooma train.[76] A Royal Air Force plane had even joined the search operations, flying in from Sydney on Saturday morning.[77] And Rutherford and the Jindabyne police were again in the field:

A party under Sergeant Carroll, including Rutherford, the black tracker from Dalgety, and seven horsemen all equipped with skis and snowshoes for use in the snow-covered country, left early to-day [Saturday] to search between the head of the Thredbo and Betts Camp, and along the Thredbo.[78]

But even as the scale of the search operation expanded, and its focus narrowed to the area near the ski marks, experience dashed expectations. A *Sunday Times* reporter described the fruitless searching in an evocative piece:

At the first streak of dawn this morning [Saturday], the Hotel was astir, and soon all preparations for the despatch of the various parties were completed. Over the snow-covered slopes little black dots scattered in different directions, and soon the rescue parties were lost in the frozen valleys. They went to the country west of the Thredbo River, down which it was believed that the men they sought to rescue would have attempted to make their way. All day long they sought among the snows – probing each drift – sliding down at breakneck speed into chill, frozen valleys – climbing laboriously up slopes of soft snow, to repeat the whole routine when they reached the top of the ridge. As the hours wore on the fire of hope, which was at its brightest this morning when the parties left the hotel, burnt lower, and when late afternoon came it was felt that only by a miracle would any trace of the missing men be found.[79]

By now optimism was waning. The headline for this report, strung across the top of the first page, ran: 'Grim Silence of the Snows Closes Round Missing Men'.

139

Even the increased aerial surveillance proved ineffectual. On Saturday afternoon one of the pilots sent a telegram to Sydney from Jindabyne:

> Carried out two hours' patrol around summit, Betts
> Camp, Perisher Range, Blue Lake, Club and Albona Lakes,
> Mt. Townsend, Ram's Head, Charlotte Pass, Merritt's
> Lookout. Altitude from 5000ft to 7000ft. ... Followed up
> all ski tracks, and examined all drifts: no success. Personal
> opinion further efforts useless. Intend returning tomorrow.
> Telegraph other instructions, if necessary.[80]

Technology had failed to provide the answer. Attempting to explain Rutherford's lack of success, the *Sun* commented that he 'is unfortunately not at all at home in the snow'.[81]

News reaching the Kosciusko Hotel increased a sense of dread concerning the missing men. The hotel manager learned that two cross-country skiers had, only days before the blizzard, been in the area where tracks were discovered. It was now assumed that these were a false lead.

That night, someone at the Kosciusko Hotel phoned Chrissie Seaman to tell her that 'no traces had been found of her husband and his companion'.[82] But while undoubtedly troubled, Chrissie did not despair easily. 'Despite discouraging reports,' The *Sunday Times* noted, 'Mrs. Seaman courageously refuses to abandon hope. She still believes her husband, with his experience of conditions in the mountains, is alive somewhere, and may be soon found by one of the search parties.'[83]

And that same evening George Aalberg returned to the hotel

with 'startling information' about the ski tracks. The false lead proved true after all.

By Saturday afternoon Evan and Laurie had been missing for four days. Not yet aware of talk that the ski marks they discovered on Friday might have been made by another party, Aalberg and Soderlund spent Saturday returning to further explore the area around these tracks. Still assuming they were the marks of Evan and Laurie, Aalberg spent much of Saturday trying to understand the trail marks, finding much more evidence to build up a picture of movement:

> He searched about the summit and a little way on the south-
> west side found tracks of a ski-er travelling alone. A little
> distance on, a second pair of tracks joined the first. He
> followed, losing them at intervals, over the southern side
> of Ethridge Range, where, instead of turning to the left in
> the direction of Bett's Camp, they bore away south to the
> Ram's Head Range. He followed them over these mountains
> and for more than half a mile on the other side, down to
> the precipitous cliffs that tower above the valley of the
> Thredbo River. The tracks led towards the roughest country
> imaginable, and zig-zagged about, showing every indication
> that someone had been endeavouring to find a way down
> over the rocky wall. In one place the marks went right up to
> the wall. Then, apparently, the ski-runner tried in different
> directions to get round, and, when baffled, turned back to
> look for another exit. Aalberg followed in and out of heavy
> timber, tangled undergrowth, and rocks, but as the tracks
> were found late in the afternoon the impressions were faint on
> the frozen snow.[84]

Aalberg returned to the Kosciusko Hotel on Saturday evening with this news. He was, the reporter recorded, 'convinced that the tracks were not those of searchers nor of expert ski-ers exploring the country, but those of men who were deliberately trying to get off the snowy ridges into the heavily timbered country below.' They looked good to be tracks of Evan and Laurie.

Before this news hit the papers on Sunday, the *Sunday Times* questioned Lennox Teece, one of the men who had reportedly been in that area before the blizzard.[85] There was one crucial piece of information that could help confirm the identity of the track-makers. Aalberg found an area where 'one skier had evidently taken off his skies, either to see whether travelling would be easier, or to work himself out of a difficult situation'. If Teece or his companion 'had taken off their skis when they were in the vicinity two days before Hayes and Seaman', then the tracks were theirs. If not, they were probably made by Evan and Laurie. When asked, Teece reinvigorated everyone's hopes:

> 'We did not,' said Dr. Teece, 'nor did we go into the timber. Undoubtedly the tracks found are those of the missing men, and if they are in the timber that is why the planes did not sight them.'[86]

With that news confirmed, the searchers again focused on that area, and a new race was on. Increasingly desperate, Aalberg co-opted a car on Sunday morning, collected Rutherford, and drove up the Thredbo valley along a bullock cart track as best he could.[87] Finally, mechanised transport was being deployed, carrying an expert skier and an expert tracker as far possible towards Kosciuszko. They were striking for the ski marks.

CHAPTER SIX

A Summit Too Far

Laurie was aware of 'the severity of the weather' he might face at Kosciuszko.[1] He mentioned as much in a letter to his parents, written prior to leaving Sydney. While most likely a generic explanation about the nature of Australia's alpine climate – Australia's winter being America's summer, after all – by Sunday 19 August it seemed an ominously prescient comment.

As Aalberg and Rutherford raced towards the site of the ski marks, Laurie's parents remained unaware of the circumstances surrounding their missing son. The letter itself mentioning his ski holiday was probably still in transit, and news of his disappearance had not reached them.

Americans were not ignorant of Kosciuszko. *The National Geographic Magazine* had devoted its December 1916 issue to 'Lonely Australia: The Unique Continent', citing Kosciuszko's 'blunt granite cap' as the country's highest point within the first few paragraphs.[2] Beside such a rote fact, the writer recognised that Australia's mountains especially determined the layout of the only nation to occupy an entire continent, stating that 'their influence is great, for nowhere is their control of rainfall and consequent distribution of vegetation and people better exemplified'.[3] In a sense, Kosciuszko was an international climatic teaching tool.

It was also a diplomatic one. In mid-1925 an American naval fleet visited Sydney to much fanfare. Notable among the many festivities was an excursion by Admiral Robinson and a party of the American officers to Kosciuszko for a much-publicised holiday with the governor of New South Wales.[4] 'They indulged in skiing and greatly enjoyed their visit' said a report on the first page of *The Guardian* in London a few days later, highlighting the international reach of news of this Kosciuszko fun.[5] Pictures of the admiral on skis in Australia were subsequently published in America, making Kosciuszko adventures a shorthand for a mutual national amity.[6]

Even the development of Kosciuszko as an everyman's holidaying space had American resonances that hinted at a new world order. Back in 1902 when one government minister was speaking of proposed developments in the area, he called for the establishment at Kosciuszko of 'a great national park, like the Yellowstone National Park in America'.[7] When a step towards that end was enacted a few years later, with a large area around Kosciuszko preserved by government, the *New York Times* noticed and reported it.[8]

But Kosciuszko could not be tamed. The Kosciusko Hotel and surrounds offered a pleasant playground for international diplomacy and the feted representatives of the mighty United States Navy stayed safely near it. Although not as high as America's highest mountains, Kosciuszko was still dangerous. The fading tracks before Rutherford and Aalberg were proof of that.

Intuitions and Exertions

By now a picture of the movements of the missing men had shifted from speculation to extrapolation. A consensus was gradually developing.

Based on the direction of the tracks, seemingly pointing towards forested valleys, Aalberg surmised that Evan and Laurie had fled from 'the severity of the blizzard' to the relative protection of 'the timbered country'.[9] He figured they probably aimed to reach 'the lower slopes below the snow line', where they could have found 'conditions warm and dry, with plenty of running water'.[10] Theoretically, he suggested, they might have been 'able to sustain themselves with roots, and, probably, rabbits'.[11] Hopes were fading, but not yet abandoned.

Others involved in the search operations shared Aalberg's line of thinking. Keith Bath, who had accompanied Laurie to Kosciuszko's summit in 1927, was 'firmly convinced that Aalberg's assumptions are correct'. In fact, Bath went further, suggesting that Laurie already had an interest in the area where the ski tracks were found:

> He [Bath] says that when they reached the summit last
> year, Mr. Seaman said: 'Why not go across to the south of
> Etheridge Hill, and pick up the road post at the bend? It will
> save quite a lot.' Mr. Bath replied, 'No! Better the snow we
> know than the snow we don't. I am going back in my own
> tracks.'[12]

Here was Laurie, the year before his misadventure, evidently keen on shortcuts.

Worried by this memory, and concerned that Evan and Laurie tried this route, potentially getting lost or going too far for a safe retreat, Bath feared that 'they are gone forever'. It was a small sign that while the searchers were still hopeful, some were starting to mourn. 'Both of them were my pals,' he added, 'and I feel the loss deeply.'

The search parties were now very fatigued. While a large party of bushmen was sent up the Thredbo River on Sunday, and three-man ski parties set out from Betts Camp to criss-cross the area of the tracks, fresh discoveries simply furthered the mystery. When he reached the ski marks with Aalberg, Rutherford concluded 'that they were probably made as late as Saturday', which was unlikely timing for men fleeing a blizzard on Kosciuszko days before that. Even allowing for some imprecision in the timing, however, the tracks told a story different from that now widely assumed. There were boot marks as well, which Rutherford thought were made 'by a man wearing a boot that was very much worn and went over at the heel'.[13] It seemed an unlikely fit for the affluent Laurie, although it was possibly a match for Evan. The man who joked about selling his shirt to go skiing could conceivably have been wearing old boots. But weighed against this tantalising lead was Rutherford's other conclusion: there was only one set of tracks.

The discovery of another set of ski tracks, in a completely different area, also seemingly heading away from the mountain, shattered the recent consensus. These were discovered by Tom Prendergast, one of the mounted parties, and he managed to follow the tracks 'for about six miles from the summit' in the upper Thredbo Valley, 'but lost them in the fog'.[14] Returning to camp, Prendergast himself then struck misfortune. 'His horse fell, and he had his leg smashed.' His companions had to carry him back to camp.

Prendergast was not alone in such rescue misfortune. Elsewhere in the snowfields, one of the skiers operating from Betts Camp sprained his ankle. The search operations were becoming increasingly hazardous.

Worse still, the weather was turning.

As Aalberg and Rutherford re-examined the tracks, the airmen had planned to converge overhead, but due to heavy winds the plan stalled:

> The two airmen, Captain King and Captain Reid, made an
> unsuccessful attempt to fly over the Thredbo Valley. High
> winds, bad visibility, and intermittent rain compelled them to
> return. Pilot Officer Carroll also made an attempt to fly over
> this country, but was forced to return shortly after starting.[15]

With so many search parties out in the field another blizzard could prove disastrous. For one party from Kiandra, it very nearly did.

The Kiandra Miners

News of Evan and Laurie being missing reached the Elaine Mine near Kiandra some days after the blizzard. In response, two brothers set off on their own rescue mission.[16] The brothers thought it possible 'that sufficient time had elapsed for them [Evan and Laurie] to have travelled even into our vicinage'. In other words, if Evan and Laurie had been forced to flee the mountain, their journey may have taken them closer to Kiandra than Kosciuszko. As such, a search from Kiandra towards the mountain was a useful addition to the main search operations near Charlotte Pass and Kosciuszko itself.

William and Robert Hughes, respectively known as Bill and Bob, were sons of 'Kiandra's skiing mailman'.[17] The pair of young men were miners. Although extensive gold mining in the Kiandra district had wound up by the 1890s, some small-scale operations had recommenced in the 1920s. The Hughes brothers worked a tunnel into hard rock, which they named the Elaine Mine in a nod to gold-mining history at Bathurst.

It seems likely that Bill and Bob were personally acquainted with Evan and perhaps also Laurie. They certainly had some mutual acquaintances. Reportedly among the major shareholders in their mining operation was Herbert Schlink, a prominent member of the Ski Club of Australia.[18] Moreover, Bill had been a local guide for a notable skiing expedition two years earlier when he led four of the Sydney skiers to successfully become 'the first to make the journey from Kiandra to Kosciusko in winter time', a journey of about 100 miles.[19] Schlink was part of this expedition, and wrote a lengthy account of it for the *Sydney Mail*, where he described Bill as 'a local resident, whose knowledge of the Kiandra area was invaluable'.[20] Also on this 1927 expedition was John Laidley, one of the men who caught the first train from Sydney to join the search operations for Evan and Laurie. It is therefore no surprise that Bill was keen to join the rescue mission.

A highly skilled skier and alpine explorer in his own right, Bill was particularly well suited to leading a search party. Fortunately, he wrote a detailed account of his preparations and movements:

> Donning the crosscountry skier's usual outfit, and including
> extra food, stimulants, etc., in our kits, we set out, entertaining
> a forlorn hope of falling in with the missing men somewhere
> south of Mount Tabletop, and intending if the weather
> continued favourable to search right through to Kosciusko.[21]

Bill's 'forlorn hope' was not simply about the time that had elapsed since the blizzard. He and Bob each had a special medal, earned several years earlier when they were part of a search party that discovered and recovered the body of a cattleman. The dead man was driving cattle when he got caught in a heavy storm.[22] Although really only boys at that point, they had seen the snow-covered cattleman and his horses and accompanied the body as it was dragged to Adaminaby on 'an improvised sledge'. They had seen what the mountain blizzards could do.

It was fine weather at the start of their search, but the recent snow provided various impediments:

> From the summit of Tabletop (5728ft.) the low country to
> the southward was subjected to a searching view through
> field glasses. About one half of the ground was snow covered,
> with bare patches somewhat regularly interspersed, giving an
> appearance of a gigantic chess board. After leaving Tabletop
> our moves towards the south were checked again and again by
> tributaries of the Happy Jack River, swollen by melting snow,
> which forced us to go many miles to the east. Towards evening
> the last stream was crossed and we turned west and reached
> the 'Booby' Hut at sunset.[23]

Melting snow caused obstructions, and patchy snow made searching difficult. The landscape was a world of contrast: snow and meltwater patched over the dull tones of grass and rock.

The going was difficult. They had been on foot thus far, but the next day they reached snow country. As they put on their skis they saw the challenge ahead of them: 'Mount Jagunal was covered by dense fog, and everywhere towards the west a dense black cloud was very much in evidence.'

This was most likely the turn in weather that impeded the aeroplanes. As it did for all the other parties searching the mountains, it changed the course of Bill and Bob's expedition. But in their case, the shift was extreme. Their mission quickly turned from rescue to survival:

The wind, which had been blowing strongly all morning, whirled fine snow around us with increased vigor in our exposed position. We struck out in what we judged to be a direct line for Gungartan Hut. The wind increased gradually in force, carrying with it fine dry hail, until some two miles from the hut it reached an intensity the like of which neither of us had in a lifetime on the snow experienced before. All thought of Hayes and Seaman had for the time gone out of our heads. We were in imminent danger of being lost ourselves, and for the time being thoughts of self-preservation were uppermost. Heads downward we plodded on. Some tracks of skis in hard ice on a ridge were examined with interest, but by the direction they were pointing and the number of them we came to the conclusion that they could not be those of the men we were searching for. ... At last the spot where Gungartan Hut was expected to be was reached, but the hut could not be seen. Shouting in each other's ears, above the roar of the wind, we discussed the advisability of running down off the high ground and spending the night somehow in the timber country. However, we decided to put in another half-

hour in search among the boulders for the hut, and to our
great relief had spent but a few minutes of this time when
we found our objective hidden behind a huge boulder and
partly covered with snow. It was found to be well stocked
with provisions and bedding.[24]

Their experience of searching for Evan and Laurie had ended up
paralleling the situation facing the missing men several days earlier.
The Hughes brothers had even contemplated running for safety
into timbered country, just as the missing skiers were now generally
assumed to have done.

The stormy conditions passed with the night, and in the morning
Bill and Bob awoke to a cold but clear day. But their experience
taught them two lessons:

It seemed to us as if any man exposed to the blizzard
overnight could not have possibly survived, and as the drifting
hail had made tracking impossible we decided to abandon
all hope of finding the missing men and return direct to
Kiandra.[25]

Evan and Laurie were probably dead, Bill and Bob concluded, and
their tracks were most likely covered. The brothers returned home,
not without difficulty.

Up towards the Kosciusko Hotel, other men were coming to
similar conclusions. One party planned to go out on Monday with
ropes, to scale down some of the cliffs and examine riverbanks and
creek beds.[26] 'It is not expected that any bodies will be found,' a
reporter noted, 'but the work is being done to complete the search
of this particular area.'[27]

Winding Down

Other searchers also turned for home. The expedition that had been photographed leaving Central Station with much fanfare a few days earlier headed back to Sydney on the afternoon of Monday 20 August. As they passed through the Kosciusko Hotel on their way home the *Sydney Morning Herald*'s 'special reporter' recorded their last day of searching up beyond Betts Camp.[28] During Sunday morning they had headed to the area around the first ski marks and the timber country. In the afternoon they followed these 'into the timber'.

> Then they opened out fanwise and fought their way over the
> precipitous slopes right down to the valley, a drop of some
> 2000 feet. They discarded their skis as they left the snow
> line, and were soon fighting their way over rocks and tangled
> undergrowth. They experienced great difficulty in finding a
> way down over the rocky walls, and at times were up to their
> waists in snow.[29]

As had been the case for the Hughes brothers further afield, this already difficult expedition was made harder by the turning weather:

> The return journey proved most arduous and exciting, and
> time and again the men could not get a foothold. At each step
> they sank deep into the snow, and experienced great difficulty
> in extricating themselves. They kept calling to each other to
> ascertain if each man was safe. The men say that they had
> no idea a mountain-side could be so rough and dangerous to
> negotiate. They report that with bad visibility the missing men

could easily have dropped out of sight in the huge snowdrifts, and even if they sought shelter alongside the rocks they would be frozen before morning. The party reached the top of the ridge again late in the afternoon, and, thoroughly tired and exhausted, they had a difficult journey back over the snow to Betts' Camp. Rain commenced to fall at 4 o'clock, and their clothes became saturated. On reaching Betts' Camp their hands were frozen and their feet sore and blistered.[30]

'It is recognised here that their feat is one of the finest ever undertaken on the heights of Kosciusko', added the reporter.

As these men packed up and began their journey home on Monday, the civil aircraft also turned for Sydney. Only a Royal Australian Air Force aeroplane continued surveillance flying. As it flew over the hotel in its final pass for the day, the pilot dropped a note over the side for the benefit of the search parties and the press. He had seen nothing of interest. The next day, Tuesday 21 August, he dropped more messages, this being the most effective means of communication available to him.[31] Despite examining the slopes and ranges with 'his military glasses' from as low a height as 200 feet, he saw nothing but the signs of bushmen and search parties.

A Story with No News

As the search efforts failed, the story started to shift. With the search missions winding down, commentary on the tragedy became more reflective and focused on contributory problems. Pointing to the lack of a telephone line between the Kosciusko Hotel and Betts Camp, commentators argued that considerable time was wasted before large search parties could be arranged.[32] This absence of rapid communication technology also meant that it was difficult to coordinate operations effectively. One member of the Millions Ski Club mentioned that he had seen 'a large number of telephone boxes' in the Yosemite Valley in the United States, clearly suggesting that such infrastructure should be contemplated in Australia.[33] Similarly, there was criticism of the government for the inadequate provision of emergency accommodation, despite regular calls for more shelters. That Betts Camp had only basic facilities was also highlighted as a problem. Apparently, it did not even have a 'proper medicine chest'.[34]

Government bureaucrats may well have been embarrassed when, with absurdly unfortunate timing, Monday's edition of the *Sun* printed a government notice that four tenders had been received for 'Installation of automatic refrigerating plant' at the Kosciusko Hotel.[35] After plugging the Kosciuszko experience so hard as a tourist experience, the government unwittingly gave the impression of being focused on refrigeration rather than rescue. And while government process could not have anticipated the alpine tragedy, the disappearance of Evan and Laurie was starting to turn from an unfortunate human tragedy into a publicity headache. The director of the Government Tourist Bureau faced a curious press when he returned to Sydney from Kosciuszko. According to the *Sun:*

> [He] adhered to the belief that there was a distinct possibility of the men having made their way into the timber country, and were perhaps still alive, subsisting upon roots, and possibly rabbits, dug out of burrows. 'The snow has been honeycombed for miles, and tracks of the missing men have been followed in the direction of the bushlands,' he added.

'Searchers have not given up hope of finding their two lost comrades.'[36]

But he had to admit that 'each day the chances of salvation grow less'.[37] Positive spin could only be maintained for so long before it looked delusional.

Some friends, however, remained guardedly optimistic. When the Alpine Club party reached Sydney, they too were met by journalists. 'If the missing men are still on the snow the outlook is hopeless now,' one said, before allowing the possibility that 'if they have been able to get down below the snow-level and into the timbered country, they would have a chance, and personally I am not inclined to think that they are still in the snow.'[38] He felt the men had got below the snowline, and were somewhere in the huge expanse at the edges of the search areas.

But while these commentators held out public hope, the lost men's families were preparing for the worst. Either Chrissie or her father let the authorities know that Laurie 'carried with him an autographic camera'.[39] This was important, a report noted, because

his relatives are anxious, if this is found at any time, that
it will not be opened or interfered with in any way, as it
is almost certain that, unless death came suddenly, Mr.
Seaman would write a last message on the film, and this, on
being developed, might solve the mystery surrounding their
disappearance.[40]

This was one of the few public mentions of the private angst of the families of the missing men. When the Seamans of Glen Cove were eventually notified of their son's disappearance by cable, the distance prevented their playing a role in the Australian press.[41] On Evan's side, his younger siblings certainly knew of his disappearance, one reaction being captured in the pages of the *Sunday Times*:

> One of the saddest men in Sydney to-day is Mr. Hayes'
> younger brother, who is employed by David Jones, Ltd.,
> Sydney. Sorrow is written all over this lad's face, and any
> mention of his brother brings tears to his eyes.[42]

But other than this small comment, and unlike Chrissie Seaman, the Hayes family played a limited public part in the unfolding story of Evan and Laurie's disappearance.

In fact, rather like the search operations, the story itself wound down relatively quickly. 'HOPE DYING', asserted the headline for the *Sydney Morning Herald*'s Tuesday article on the Kosciuszko misadventure.[43] While Wednesday's report claimed that 'SEARCH CONTINUES', Thursday's edition followed with 'Fruitless Search'.[44] On Friday it was 'STILL NO NEWS'.[45] The articles became smaller, as there was scant new information to report. Bushmen, police and some Millions Ski Club members continued to search even though the weather deteriorated over the week. One report mentioned it was becoming 'extremely cold, and it is raining heavily', further demoralising the search parties. By Monday of the following week, 27 August, the headlines took a final turn: 'SEARCH ENDS. Kosciusko Tragedy. PATROL FOR BODIES.'[46] The new consensus was that bodies would be found in the summertime once the snow had melted away.

In the final week of search operations at Kosciuszko, one reporter obviously tried to reinvigorate the story with a mystical twist. He found a woman at Jindabyne who claimed to have

had a vision 'whilst sitting nursing her baby by the fire'.[47] Hilda
Coleman described seeing 'two men – one lying dead in the snow
near Smiggin's Holes, and another sitting under a rock near
Charlotte Pass'.[48] Thinking that 'Mrs Coleman seemed greatly
affected by her dream', this reporter interviewed her for further
information:

> 'I saw the two men in the vision as clearly as if I was gazing
> at them in the flesh,' she said, 'and I am convinced that
> some unknown power is using me as a medium to guide the
> searchers. The man sitting under the rock had his head bent
> down, and he was holding what appeared to be a camera or
> a book. The other man was lying half-buried in the snow. He
> was wearing dark trousers and a pull-over sweater.'[49]

But as a news item, this vision got only a little traction in some rural
newspapers, seemingly too silly for the major metropolitan dailies.[50]
Even a supposedly 'extraordinary coincidence' of 'the ominous
gathering of a circling body of crows high over the spot where
Mrs Coleman saw the body in her weird dream', proved nothing
in the end. Most likely the wider press felt that the coincidences
between the dream and published facts and opinions were too good
to be true. Its timing and content seems to have suspiciously aligned
with reports of Laurie's camera and widespread acceptance that the
men were probably dead and snow-covered.

As one of the major newspapers to have covered the story
seriously from start to finish, the *Sydney Morning Herald* reflected
on the tragedy in a lengthy editorial. After describing the search
operations, and admitting the likelihood that Evan and Laurie were
dead, it discussed their fate:

It is consoling to reflect, however, that death from such a
form comes mercifully. Men who have narrowly escaped
being overtaken by a like fate and have been rescued when
actually unconscious, assert that there is no pain. The
victim is not even sensible of his impending doom. He feels
an increasing lassitude and drowsiness, and eventually
passes over the border-land as though through the gates of
slumber.[51]

In the days that followed, life at Kosciuszko gradually returned to
normal. In the last week of August 1928 the *Manaro Mercury, and
Cooma and Bombala Advertiser* reported that 'numbers of school
pupils arrived by train for Mt. Kosciusko'.[52] Australia's winter
playground was back to its usual business, at least on the surface.
But it would not stay that way for long.

A Time for Prayer

During the crisis, the friends and family of Evan and Laurie turned
to prayer. As the *Sunday Times* noted:

Mass will be held for Mr. Hayes, the missing skier, at
The Sacred Heart Church, Randwick, on Sunday next [9
September]. The Service has been arranged by a lady friend
of Mr. Hayes, who greatly admired the popular sportsman,
although she only met him on very few occasions, and then on
a purely business footing.[53]

It was a touching gift for Evan, recalled in the same announcement
as 'one of Sydney's best-dressed young men and apparently one of
the most popular'.

The Mass was scheduled to coincide with what the Catholic *Freeman's Journal* described as 'The Greatest Sunday in Australian History'.[54] With banners, benediction, huge crowds, hymns and even 'Maoris in full dress', a large Eucharistic Procession concluded the International Eucharistic Congress being held in Sydney. The spectacular show of Catholic devotion was considered a tremendous success, at least in Catholic circles. Not only had the faithful come out in large numbers, but non-Catholics had too. 'Never before has Australia seen such a demonstration of religious toleration,' the *Freeman's Journal* enthused, 'and the hearts of all Catholics and non-Catholics filled with pride at the beautiful display of good citizenship and decent feeling.' Removed from these much-recorded displays in the city, Evan Hayes was remembered in a quiet suburban church. But the story of this Mass was soon forgotten, drowned out by the noise of the Eucharistic Congress and fresh news from the mountain.

Initialled W. L. S.

The first hard evidence of Evan and Laurie's fate was found on the afternoon of that prayerful Sunday in Sydney. Having summitted Kosciuszko with one of their teachers, Evan Mander Jones, a party of Church of England Grammar boys were heading back down when someone spotted something a little off the main trail. Closer investigation followed. Laying a few hundred yards away were 'two skis with their stocks lying in them'.[55] The equipment had obviously been exposed by melting snow. Intrigued, and very likely aware what the find may have signified, the teacher examined them closely. He soon 'saw they had the initials W. L. S. branded near the toes'. The initials stood for William Laurie Seaman. After all the searching,

Laurie's skis had been discovered on the route towards Betts Camp, only about a mile or so from the summit of Kosciuszko.

Jones now looked closely around the immediate area. Some dozen or so yards from the skis, near 'a heap of low rocks', he found a snow-covered body.

Noticing 'a camera in a case slung over one of the shoulders', Jones 'pulled the camera out of the case and read the name "W. Laurie Seaman" impressed thereon'.[56] He put the camera back in its case, took note of his surroundings, and continued with the boys to the Kosciusko Hotel, where he promptly informed the hotel manager of the development.

By the time Jones reached the hotel it was too late for a retrieval party to go out. Laurie had to spend one more night in the snow.

In the meantime, the manager made some phone calls. He notified the Jindabyne police at about 10.30 pm, and at some point during the night discussed the situation with the director of the Government Tourist Bureau. It would also seem likely that Chrissie was informed that Laurie had been found.

In the morning a small team headed out. The recovery party included veterans of the search operations like Constable Lambert and George Aalberg. Once more they fought through gale winds and sleet looking for Laurie, but this time Jones could lead them straight to the body. Laurie was:

lying on [his] back with one hand straight down [his] right
side, the left hand was stretched at right angles to the body
and the feet were crossed. The head was inclined slightly to
the left side; he had glasses and a cap on, but no gloves.[57]

After briefly staring at Laurie laying inert in the snow, they hurried him back to the hotel. The next day the local coroner held an inquest, taking evidence from several key figures. A medical doctor examined Laurie and concluded 'that death was due to exposure to extreme cold'. Laurie showed no signs of major injury.

Laurie's father-in-law travelled up by train and attended the inquest. He took the opportunity to address the proceedings:

> Mr. Ernest W. Bell, father of Mrs. W. L. Seaman, who was present at the inquest, stated that he had been asked by Mrs. Seaman to express in open court her heartfelt gratitude to all who had so nobly assisted in the search for the bodies of her husband and Mr. Evan Hayes. That lady also expressed her gratitude and appreciation of all that had been done in the occurrence.[58]

The coroner responded in kind, publicly thanking the people of the district for assisting in the search. He also asked Bell 'to convey to Mrs. Seaman and relatives the sympathy of all concerned in the mountain search. It might be comforting for them to know that the indications were that the late Mr. Seaman had passed away without mental or physical suffering', he added.

From Sydney, Chrissie also thanked those connected with the search for Laurie. She wrote individual letters of thanks to certain key people, including the police tracker:

> Dear Constable Rutherford, – I feel at a complete loss to convey even in a small way my great gratitude to you and the many others whose brave and untiring efforts meant so much to me and mine in the great loss that has befallen me. Words

at a time like this are so futile. With my life's memories will be linked now the remembrance of your noble and unselfish deeds in the search for my dear, dear husband. In my deep sorrow I cannot adequately express my thoughts, but, oh! please do accept my heartfelt thanks. – Yours very sincerely, Chrissie Seaman.[59]

Chrissie's grief was clearly profound. She may well have especially felt the loss of her husband at this time, because by then she would have known herself to be pregnant.

A Funeral

In the *Sydney Morning Herald*'s Tuesday death notices, between a 'dearly beloved husband' and a 'beloved mother', were three notices for Laurie.[60] The first described him as 'beloved husband of Chrissie Seaman', the second as an 'only son', and the third as both 'son-in-law' and 'partner'. All ended with the same sentence: 'Perished in the snow at the Summit, Kosciusko.'

After the coronial inquest, Laurie was conveyed to Cooma and 'placed in St. Paul's church'.[61] There he lay while awaiting the north-bound Cooma Mail train, which carried him down to Sydney.[62] Once there, he was prepared for eternity at 'Kinsella's private mortuary chapel' in Oxford Street. Thursday's edition of the *Sydney Morning Herald* advertised that Laurie's funeral was due to commence at the chapel at 2.15 pm that day.[63]

'Touching scenes' reportedly marked the service.[64] The pallbearers were members of the Millions Club and the American Legion, both highlighting Laurie's international connections. The chapel was 'bedecked with flowers', including many floral tributes:

Among the large number of wreaths were those from 'The Veitchs,' 'The Bell Service Club,' principals and staff of McLachlan, Westgarth and Co., Kosciusko Alpine Club, president and members of the Millions Club, directors and staff of the London and Lancashire Insurance Co., Art Furnishings Ltd., members of the B.I.E.S.A., rector and parishioners of Cooma parish, Sun Insurance Co., Sir Charles and Lady Rosenthal, Messrs. H. J. Lamble (director of the Tourist Bureau), and M. T. Sadler (manager Yorkshire Insurance Co.).[65]

Another wreath was from 'The brothers of Evan Hayes'.

Attendees were numerous, itemised in the newspapers much like the wreath-givers.[66] Besides the family, notable people included the consul-general of the United States and, much further down the list, 'G. Hayes'. Also present were figures personally connected with the search missions like Sodersteen, Douglas and Soderlund.

Reverend Richard Ord Todd conducted the funeral service according to the rites of the Church of England and, befitting the former serviceman, an army bugler played the 'Last Post'.

Outside, many members of the public gathered to pay their respects. Clearly Laurie's story had somehow touched their own lives. 'Several hundred people,' one report mentioned, 'mostly women, stood with bowed heads as the cortege left for the Church of England portion of the Rookwood Cemetery.'[67] One woman fainted.

Death Imagined

Up in the mountains, a fresh search for Evan was being made around the area where Laurie was found. But it was unsuccessful, leaving his whereabouts a mystery. In the absence of evidence, some members of the public turned to their imaginations to fill in the gaps, revealing how the Kosciuszko tragedy affected the emotional life of the nation.

Back when a consensus gradually formed that Evan and Laurie were most likely dead, the *Sydney Morning Herald*'s concluding editorial grappled with the peculiar interest the story had generated, and the role of death within it:

> The impression created by this tragic affair is the more
> profound because of the unusual character of the occurrence.
> Constantly we hear of persons being killed in motor accidents.
> These fatalities excite sympathy and regret, but they are
> commonplaces: they are part of the risk entailed in living in
> an age of swift mechanical transport. The Kosciusko tragedy,
> however, is on a different plane. It is something foreign to our
> experience.[68]

The gist of the newspaper's argument was that there was a popular misconception that Australia was universally dry and brown and flat, and that a freezing death did not match common stereotypes. But by detailing the nature of a freezing death, it also underlined the fact that this was something quite foreign to many of its readers' daily experiences. Most Australians did not risk dying from exposure in the snow.

Another newspaper went further, delving into these presumed mountain deaths in a bizarre fashion in its final August edition for 1928. *Smith's Weekly*, which subtitled itself 'The Public Guardian', was more of a light-entertainment magazine than serious newspaper.

Death seemed to be a running theme for that issue. The leading story covered a Sydney gardener lingering near death for several months, possibly due to plant poisoning.[69] Complementing it was a cartoon titled 'Drawings of Australian Life – No. 2: When Strawberry was run over', depicting a dead cow surrounded by a crowd. Two pages further was 'The White Death', which was an 'Imaginative Reconstruction of a Tragedy'.[70] It was obviously inspired by the fact that 'KOSCIUSKO CLAIMS TWO LIVES', as the banner announced. Without naming Evan or Laurie, it opened as follows:

> How it happened he really couldn't say. One moment it was all
> plain sailing, and then suddenly he was spinning through space
> to fall in a huddled mass of pain in a fold in the mountain
> side. Both skis had been smashed from his feet, and one ankle
> was shot through and through with needles of fire. Then the
> blizzard wrapping the snow high about him in a merciful
> embrace of warmth. Next dark and the stars – lamps of yellow
> fire hung against the blue of the inverted bowl of night.[71]

Short stories of such imaginative fiction inspired by real events were commonplace. After so many young men had died overseas in the war, Australian writers often turned to fiction to imagine their final moments and thoughts, and so in one sense it was natural that the same habit could be extended to the tragedy on Kosciuszko.

The story gave the suffering skier a voice and memories, drawing on a common cultural milieu:

> There is one sound only – the beating of his own heart. 'AND
> ALL THE AIR A SOLEMN STILLNESS HOLDS ...' His
> mind harks back to that odd scrap of verse of a forgotten
> boyhood. Forgotten, that is, until this night, when memories
> come crowding ... unbidden but insistent memories.[72]

After thus referring to Thomas Gray's 'Elegy Written in a Country Churchyard', the imagined skier muses upon the nature of silence for some time, comes back to the moment by focusing on their own pain and discomfort, then gradually seems to slip into a mad dreaming state where he talks about fires and food. He then briefly returns to the poem again at the end of the piece:

> 'And all the air … sometimes stillness holds.' Funny that
> should come back. Knew the whole poem once. Learnt it for
> old Smithy when he had us for English in Fourth form. Dead
> nuts on talkin' in class, Smithy. He'd do well down here.
> Look! There he is coming down behind the Prep. school.
> 'Right oh, sir! Coming right away…'[73]

It was perhaps a measure of the sort of collective grief that people felt towards the victims.

The events at Kosciuszko also inspired the recitation of memories of those who had faced similar dangers in the past. Just as fiction had been called upon to interpret the present, so too was history tapped as an aid to understanding. In Armidale in northern New South Wales, a jeweller named Alleine Fletcher recounted his experience as 'an adventurous youth' on Kosciuszko back in 1888 when he was 'overtaken by a blizzard or snow-storm'.[74] Fletcher and a companion lost their pack-horse in a sudden change in weather, took an hour to re-locate it, and then 'fell over the gorge, pack-horse and all, being caught on a ledge 30 feet below'.[75] Escaping this situation with much difficulty, they endured a terrible time over several days before finding safety.[76] While the memory spoke of the possibilities of survival, by the time it was published hopes for Evan and Laurie were rapidly fading.

As well as being cues to interpretation, such historic and fictive stories help reveal the widespread interest in the story. In fact, while the events at Kosciuszko are best reconstructed from those reports closest to the action, readers from Western Australia to New Zealand also followed the developing story through cable reports in their local papers, even if stripped of much of the detail. But while it was the newspapers of Sydney and those closest to the mountain that showed the greatest interest, there was one notable geographical exception. For readers of the *Glen Cove Echo*, this was a story about a lost American son.

From America to Australia

About a week after William H. Seaman was first informed of Laurie's disappearance by cable, the story was front page news in the *Glen Cove Echo*, where Laurie's father was the city mayor. Under the heading 'W. L. SEAMAN LOST IN STORM IN AUSTRALIA', the blizzard and search operations were briefly described.[77] 'Mayor and Mrs. Seaman are arranging to sail for Australia', the article added.

> After consulting with the members of the City Council and
> his colleagues on the Board of Supervisors, all of whom
> agreed that it was best for him to make the trip, he is making
> preparations to sail from Vancouver on Sept. 15th, leaving
> Glen Cove about the 5th.[78]

In America, this parental mission became the main angle of the story. The *Brooklyn Daily Eagle*, for instance, carried the news under the title 'Glen Cove Mayor to Seek Son, Missing in Australian Alps'.[79] But as things panned out, the Seamans were still in Glen Cove on 11 September 1928 when, as a special report for *The New York Times* detailed, they received confirmation that Laurie was dead.[80]

After travelling to Vancouver in western Canada, the Seamans boarded the *Niagara*. It was bound for Australia via intermediate Pacific Ocean ports and departed with much fanfare. Returning members of the Australian Scottish Delegation were aboard, and they reportedly 'sang favourite old Scottish songs' with the large crowd at the wharf joining in.[81] These Australians of Scottish heritage had been travelling through Britain and Canada since August, part of an exercise in strengthening cultural and economic connections within the British empire. Also aboard was a victorious New Zealand bowling team, the Under-Secretary of State for the Dominions, and '3,700 tons of apples, onions, salmon, and lumber'.[82]

Among the 479 passengers on the liner, the Seamans mostly kept the purpose of their journey quiet. Writing home from the ship to his deputy, William Seaman explained:

> we have thought it wise not to tell our fellow passengers our mission, excepting two personal friends that we have met who are going back to Sydney after six months' absence.[83]

It seems that they wanted to keep their visit from becoming a spectacle.

As a shipboard routine developed, one incident likely caught the Seamans' attention. When the ship was heading towards Hawaii one of the Australian Scottish delegates, 72-year-old Sydney Arthur Mott, attended an evening concert on the main deck, probably in the company of a niece from Albury, who was travelling with him.[84] Only the next morning was the alarm raised. It appeared that during the night 'he accidentally fell overboard through a sudden lurch of the ship'.[85]

Having lost one passenger, the *Niagara* gained a notable one at its next port. Sir Joseph Carruthers boarded the ship at Honolulu, having been there as an Australian representative to the joint American–British commemoration of the death of Captain James Cook. The former Premier of New South Wales brought aboard tales of the emotional ceremonies that, he later asserted, 'did more to bring the people of America and of our Empire into closer touch than any other formal meetings of which I have heard or have had experience'.[86] With its gathering of warships from Australia, New Zealand, Japan, the United States and other countries, the laying of a commemorative tablet, replete with the sounding of the 'Last Post', was, Carruthers said, 'memorable beyond my powers of description'.[87]

The rest of the voyage was less eventful, although William Seaman took advantage of the *Niagara*'s arrival in Suva in Fiji to post his letter home. As he explained to his deputy, he was still rather in the dark about the full details and sequence of events concerning Laurie's disappearance and death:

> I have been up twice talking to the wireless man to find out
> the name of the other man lost with Laurie, but he could
> only remember the name of Seaman ... We are still anxiously
> awaiting the details.[88]

By this time the *Niagara* was accelerating its schedule, as there were plans to convert one of the holds into 'a refrigerating chamber' when it reached Sydney. The Seamans' wait for information was therefore diminished a little.[89] After stopping at Auckland the *Niagara* reached Sydney on the afternoon of Friday 12 October.[90] By then the Kosciuszko story had completely disappeared from the headlines and the Seamans disembarked unannounced.

Pilgrimage

It rained at Kosciuszko on the day that the Seamans reached Sydney, but their movements largely remain unclear because their time in Australia was relatively undocumented.[91] Presumably they stayed with their daughter-in-law and her father, doted on their grandchild Bruce Laurie, and visited the grave of their son. Before they left Australia, they journeyed to Kosciuszko to the site where Laurie was found. They also met with some of the men who last saw him, as well as many of his friends and acquaintances around Sydney. Once again, their visit was briefly described in a letter William Seaman sent while he and his wife were journeying home to Glen Cove:

> Laurie had made many warm friends in Sydney during his
> three years and all assistance possible was given for him and
> his friend that could have been done anywhere. All the people
> were most kind possible. The body of his friend, Evan Hayes,
> was still unfound at the time we left on November 17th. I was
> fortunate enough to get up to the spot where Laurie's body
> was found and interview the young men who last saw Hayes
> and Laurie, many of the searchers and the young man who
> found Laurie's body. Most of their doings, after they were
> last seen, will be a mystery. But the location and condition
> of Laurie's body was comforting. I hope they soon find Mr
> Hayes's body, which they hoped to do during December.[92]

But while much of the story of what happened to Evan and Laurie remained a mystery, by the time the Seamans reached Australia one important clue had been recovered: the film in Laurie's camera.

CHAPTER SEVEN

A Vital Clue

Images of Evan and Laurie remain preserved in newspapers, relics of the reportage their disappearance generated. They spoke of recent worries and happy past seasons. Among them was proof of Laurie's first journey to Kosciuszko's summit in 1927. In it, Laurie stands with two companions at the top of Australia, staring out of the grainy scene towards the camera.[1]

Such Kosciuszko photography was a familiar phenomenon. And that Laurie would take his own camera into the world that was opening up on Kosciuszko was not in the least bit surprising.

In that same season when Laurie summitted Kosciuszko, some of Evan's own adventures in the snow were printed in Sydney's *Daily Telegraph*.[2] In one photo, he sits on the snow with two mates, somewhere out in the fields. In another, he is caught skiing 'up on to the roof of the engine house of the Hotel Kosciusko', as the caption explained, in company with 'Miss Dulcie Ibbotson'. Achievement, adventure, friendship, courtship. Kosciuszko had it all, and it was there for all to see.

Images of snow, skiing and sled-riding often filled the pages of Australia's newspapers in these years. The peaks, rivers and valleys attracted the ready attention of photographers and editors, but more often it was people that drew the photographer's gaze. The 8 June 1927 cover of the *Daily Telegraph*, for instance, was mostly taken up by a picture of young Lucy Cameron skiing earnestly, while a double-page spread within showed young and old play-fighting, competing and smiling.[3] Later in the month, schoolchildren waved and fought in a fresh set of images. 'Youth runs riot at Kosciusko', one caption quipped.[4]

A month later, George Aalberg dominated another of the *Daily Telegraph*'s front pages, his arms high and his feet far from the ground. The image was taken part-way through a '75-feet jump'.[5] Only three days previously Aalberg took up five photos in another of the *Daily Telegraph*'s spreads of Kosciuszko shots, exhibiting jumps and swerves to the obvious interest of at least one photographer.[6] Not to be outdone, the *Sydney Mail*'s first August edition that year showed Aalberg from an even more dramatic angle, part-way through 'A Jump of 80ft at Kosciusko'.[7] Even before the tragedy, Kosciuszko was part of the very imagery of Australia.

Collectively, these images spoke of the way that Australians liked to see themselves. Besides the obvious sportiness and competitiveness, the photographs spoke of fashion, especially when women were the main subject.[8] Between gum trees and the hotel, the imagery also blended the distinctly Australian environment with hints of European culture. And while adults featured often, schoolchildren probably did more so, their ski-clad adventures the very image of a nation of youth and promise.

But what Laurie Seaman's camera preserved proved surprising.

The Camera

After being recovered with his body, Laurie's camera was handed to Chrissie. She then 'entrusted the work of developing the films still in the camera to Mr. Bradley, of the Melba Studios'.[9] As revealed when Laurie was still missing, Chrissie was hoping that he might have used the camera to send her a final message, somehow recorded in film.

The Melba Studios had been in Sydney since at least August 1911 when an advertisement alerted 'Newly Married Couples' that they had 'the best lighted studios in Sydney' and were 'now open'.[10] Prominent in the history of photographic portraiture in Australia, Melba Studios had already opened in Melbourne, where by 1897 visitors to the city were advised by one writer to 'lose no time in seeing for yourself the elegant rooms situated at 101 Swanston-street, which are reached by a lift'.[11] In 1898 there was even a Melba Studios office in Kalgoorlie, ostensibly the 'LARGEST and BEST LIGHTED STUDIOS on the Goldfields'.[12] But with the advent of cheaper, more portable hand-held cameras like Laurie's, the importance of the studio had shifted from portraiture to include development. Seeing as Laurie's camera had endured more than the safe sterility of a photographic studio, their expertise was most likely to get results.

'The films were in a bad condition,' a reporter noted, 'having been saturated in the snow for a month. The paper and backing of the film were pulpy.'[13] But against the odds 'Mr. Bradley managed to secure very fair prints'.[14] From the chemistry of the photographic studio two images resolved out of the darkness. Each showed a man standing at the summit of Kosciuszko.

These were the last recorded moments of the lives of Evan Hayes and Laurie Seaman. But they clarified a much bigger picture, bringing important evidence to bear on the sequence of events. After waving goodbye to their friends at Charlotte Pass, the pair had in fact continued to the summit of Kosciuszko. After climbing the mountain, Laurie took his camera out of its case, and he and Evan did what thousands of Australians and visitors to Australia have done. They posed by the cairn and photographed each other at Australia's highest point.

When the photographs were recovered the *Sydney Morning Herald* again got the scoop, publishing the images and the story of their recovery. As it explained, the photographs offered clues beyond the simple fact of reaching the summit:

> Mr. Laurie Seaman is taken leaning on the obelisk which marks the summit. While being snapped he lifted his snow goggles up on to his forehead, and is shown without his skis. Mr. Seaman then took a photo of Mr. Evan Hayes, standing alongside his skis. He is shown well muffled up around the neck. The photo was evidently taken shortly after noon, and … help materially to piece together the story of how the two men came to be lost. The body of Mr. Seaman was found about 50 yards off the regular summit track, and less than a mile from the summit. George Aalberg … found footprints leading to the summit, and the single ski marks leading away from it. This evidently shows that Mr. Seaman left his skis and gloves where they were found near his body, and walked over the hard glassy snow to the summit, accompanied by Mr. Evan Hayes, who preferred to use his skis. Seaman, in returning, kept to the road; while Evan Hayes went over the hills to the left of Etheridge Range to cut out the horseshoe

bend in the road, and evidently got lost as the blizzard struck him. ... Evidently, while waiting for Hayes to return Seaman himself caught the severity of the blizzard. He was so near the main track that is it not unreasonable to assume that he also lost his bearings. His body was found in a sitting position, with his head leaning up against a rock. If the late Evan Hayes was lost before rejoining Laurie Seaman – and it looks like it – there is no saying where his body will be found. He probably wandered about for some time hopelessly lost.[15]

In this reconstruction of events – still rich with supposition, and a perhaps misplaced understanding of how proximate Laurie's body was to the summit, as abandoning skis at that distance would be rather strange – Evan and Laurie took separate routes down from the summit and planned to meet where Laurie had left his skis. Reaching the spot alone, Laurie sat down to wait for his friend, who never arrived.

Beyond simply providing some clues to the sequence of events – which probably read too much into the fact that Laurie's skis were not in his portrait – the definitive proof that Evan and Laurie had ascended Kosciuszko provided a much clearer sense of where Evan might have gone. With the coming of summer, his friends once more set out to find him.

Seeking Evan

In early December 1928, two of Evan's friends led an expedition from Sydney to the Snowy Mountains 'to search the snow slopes of Kosciusko for the body of Evan Hayes'.[16] Keith Bath of the Millions Club, together with 'the well-known skiing expert' Francis

Austin Parle, organised the party of about twenty men. Bath was an alderman of the Sydney beachside suburb of Manly, and Parle was a chemist at Campsie. Like Laurie, Parle was a motoring enthusiast, and had won one of the reliability tests to Avon Dam earlier in the year in his Bugatti.[17] He was one of the men who first took the Cooma Mail train to join the search and rescue operations after the blizzard. Bath was a well-regarded and active skier, who had summitted Mount Kosciuszko with Laurie in 1927.

The plan was straightforward, although it involved considerable logistical arrangements:

A three or four days' search will be made, particularly in the locality where it is surmised the unfortunate skier was lost.
The party will traverse the country on foot, and ponies, as the snow at this time of the year is not deep, and it is thought that the body if still in the snow, will now be partly uncovered.
The men will search in extended order, so as completely to comb the hills and valleys. The men will be conveyed by car as far as Charlotte's Pass, about four miles above the hotel, and they will work out from this centre.[18]

Once in the mountains, the expedition was bolstered with the assistance of 'bushmen from Jindabyne', but even after several days of determined searching in 'the most likely localities', Evan remained missing.

The expeditioners returned home, doubtlessly a bit dejected. As the *Sydney Morning Herald* reported:

after seeing the nature of the country in the summer time, [Bath] came to the conclusion that the task, even with a party of 50, would be a hopeless one, as the country is very rugged, and consists mainly of outcrops of boulders, 15 to 20 feet high. He states that even with men a few yards apart, it is almost impossible thoroughly to inspect the insurmountable crevices between the rocks.[19]

Instead, he put his hope in the pastoral industry:

He points out that shepherds, who are pasturing flocks of sheep are gradually making their way up to the higher levels now that the snow has melted, and these men during the summer season are the most likely to succeed in the search.[20]

To encourage any visitor to the mountains to keep an eye out for their lost comrade, Bath offered 'a reward of £20 for the recovery of the body'.[21]

Even before the offer of money, Evan's friends were not the only ones looking for him. Immediately following Bath and Parle was another police expedition, aided by the resources of the Government Tourist Bureau and personnel from the Kosciusko Hotel. With them was 'a tracker attached to Dalgety police', presumably Rutherford.[22] Setting out on 19 December, this party soon also encountered difficult conditions:

Etheridge Ridge, Dead Horse Ridge, and the country between the latter and Merritt's Camp on the Thredbo Falls, were thoroughly searched ... Particular care was taken to search the country thoroughly within a circle of a mile from where the body of L. Seaman, who disappeared with Hayes, was found. The rugged nature of the country made searching difficult. The ridges in places were a jumble of loose rocks, sometimes so rough that on the second day of the search the horses were abandoned and the search was continued on foot.[23]

After five days of such searching, no body was discovered. The police planned further operations, but their lack of success meant they went under-reported.

But summers in Kosciuszko country were increasingly busy, so there were hopes for an incidental find. Aside from tourists, the mountains had long had a rhythm of summer grazing going well back into colonial days. In the 1920s descendants of both people and cattle still wandered the grassy slopes at certain times of the year. As a correspondent explained to the Melbourne *Age* that December:

> A number of graziers are removing their store bullocks back into the mountainous country towards Toolong, N.S.W., and the spurs and gorges of Kosciusko, where there is remarkable summer feed, knee high and green. This is the usual procedure at this time of year. When the snow melts and partially disappears the summer grasses take its place, and cattle are 'topped off' there until February or March, when they are trucked to Melbourne markets.[24]

Like the police, the cattlemen appear to have found no sign of Evan.

Neither had the shepherds, and there were plenty of them too. In January 1929 the front page of the *Evening News* carried a photograph of sheep in the Snowy Mountains. Its caption explained:

> Great flocks of sheep are being driven from the drought-stricken areas in the country to well-watered uplands of Kosciusko for relief. There is abundance of water and grass on the high ranges.[25]

The numbers of sheep being driven to the mountains from the Riverina and the Monaro were astounding. Back in the summer of 1877 a correspondent observed that 'there may be found located some 300,000 sheep in every imaginable nook and corner from "Long Plain" to "Happy Jack," and reaching on to the remote Mount Kosciusko and the boundary of the Snowy River'.[26] Since those days the numbers had certainly increased. One number given in February 1929 for sheep using a road into the snow country had 1,065,000 of them grazing their way up from Cooma.[27] This was not even the extent of it. More flocks converged on Kosciuszko from different directions.

But the sheep proved ineffective search agents. Summer turned to autumn and the chances of finding Evan again diminished. 'As the sheep pastured on the highlands in the warmer weather have now been mustered and taken to lower levels,' the *Sydney Morning Herald* noted in April 1929, 'all hopes of recovering the body of Mr. Evan Hayes must be abandoned for several months.'[28] But there had been progress of a different sort on the mountains. 'The Laurie Seaman Memorial Hut', the paper reported in the same article, 'is almost completed.'[29]

Memorialising Laurie

Before departing Australia in November 1928, Laurie's father 'wrote to the ski clubs who took part in the search, expressing his deep gratitude to the members for their untiring efforts to locate the missing skiers'.[30] He also made preliminary arrangements for the erection of a memorial to his son at Kosciuszko.

William Seaman donated £150 and left the planning and execution of the memorial to a committee. This was comprised of representatives of three major ski clubs who would work in consultation with Laurie's father-in-law.[31] The planned memorial was going to be a practical contribution to the Australian alps, in the form of a much-needed 'shelter hut' built on the range where Laurie had died.

While the hut was being planned and built over the warmer months, Kosciuszko was inundated with its usual complement of summer tourists. In addition to the cattlemen and shepherds, drivers and walkers, fishermen frequented the Snowy and Murrumbidgee Rivers and their tributaries seeking trout. This was all part of that ongoing government push to make the area attractive for recreational use. As early as 1906 the New South Wales government announced a scheme to effectively create a game reserve around Mount Kosciuszko, which they planned to stock with deer.[32] Unsurprisingly, the deer idea did not stick. But the government followed through more effectively with trout. In 1925 the government's Fisheries Department stocked the alpine lakes about Kosciuszko in preparation for recreational fishing over the coming seasons:

> The fish (which were 6½ months old) were conveyed from
> the hatcheries at Prospect in cans, and distributed, with the
> assistance of pack horses, at the summit of Mt. Kosciusko in
> Lakes Cootapatamba and Albina, and in the headwaters of
> the Snowy River.[33]

In 1929 one angler went so far as to claim that 'there are few districts in the universe where so many streams provide, in small area, such wonderful trout fishing as may be enjoyed, free for the taking, on Kosciusko's sides'.[34]

The quiet fishermen may, however, have been occasionally disrupted. That summer a complement of 'the Harley Motor-Cycle Club' from Melbourne rode into Kosciuszko.[35] Cooma's *Manaro Mercury* noted with obvious interest that of the thirty-three members, nine were women. 'One lady', it mentioned, 'drives her mother in the side car, and is an adept in motor cycling.'[36] Undoubtedly these touring groups struggled at certain points of the road to navigate flocks and herds.

Another tourist, who also happened to be a journalist, was surprised to meet someone he knew while taking a driving holiday in the mountains in late March. 'Whilst motoring in the vicinity of the Hotel Kosciusko', he said, 'a man that was standing on the roadside held up his hand and called out.'[37] This fellow was from Port Kembla and explained that he was in the mountains building Seaman's Hut.

By early April 1929 this memorial to Laurie was 'almost completed'.[38] By mid-May it was finished:

Six thousand eight hundred feet above sea level, the chalet is visible for miles around the mountain tops. Dominating the snow-capped top of Etheridge Ridge, which stands out conspicuously, the chalet is in a most handy position for skiers and walking parties. The chalet is within a few yards of the rocky outcrop where the body of Laurie Seaman was found. The building comprises two rooms with an entrance porch to afford protection and for fuel storage purposes. The walls are of granite rubble, backed with concrete with an internal surface of insulating material. A fuel stove is provided, also a cupboard for storing food. The key of the door is placed in a receptacle in the exterior hall.[39]

The hut was not merely an additional structure in the alpine landscape. It also encapsulated new approaches to alpine tourism, learned from the tragedy of 1928. For instance, it was decided that Seaman Hut should be connected to Betts Camp by telephone lines.

The ceremonial opening took place on 17 May 1929. Among the attendees were representatives of the Millions Ski Club, the Aero Club of New South Wales, the Royal Automobile Club of Australia, and other such organisations with various connections to the Seamans or the tragedy. Dignitaries included the chief secretary of New South Wales, the consul-general of the United States, the director of the Government Tourist Bureau, and the manager of the Kosciusko Hotel. Emil Sodersteen was also present, representing those who had last seen Evan and Laurie.[40]

But the opening day did not start smoothly. The trek to the hut proved a troublesome journey itself, as the *Sydney Morning Herald*'s 'special representative' related:

> The sun was shining brightly when the party, numbering
> 20 or more ... set out for Deadhorse Ridge on the Etheridge
> Range. A small fleet of cars moved warily over a slippery road
> that resembled a quagmire. An unusually heavy fall of snow
> during the preceding two or three days had begun to thaw.
> Over patches of loose earth, the cars lurched dangerously.
> The drivers were the objects of unqualified admiration, but
> pessimistic members of the party rightly anticipated the
> unspoken fears of the remainder.[41]

The lead car got bogged and had to be righted by the crowd before the convoy could continue. Eventually the snow got too thick and mushy, and 'the cars could go no further'. The party had to walk the last five miles, which quickly 'revealed the inexperienced mountaineers' among the group.

The difficulties of the climb, the reporter thought, helped foster a sense of 'camaraderie that was sincere, and lasting'. He gave one pointed international example:

> Two or three miles had been left behind when the United
> States Consul-General (Mr. Lawton), who was trudging
> bravely on to represent Mr. Seaman's parents at the dedication
> ceremony, had a consultation with the Minister [the Chief
> Secretary], and then sat down. 'What!' called out one of the
> experienced alpinists surprisedly, 'the Stars and Stripes! Do
> they ever say die?' 'No,' Mr. Lawton replied; 'they never say
> die,' and, with a determined glint in his eyes, resumed the
> march.[42]

Eventually the party of men and women reached the hut. There, Chief Secretary Frank Augustus Chaffey oversaw the short ceremony, which obviously touched the special correspondent who witnessed the event:

> In a weird stillness, a harrowing, raw wind, with the
> thermometer registering freezing point, and gigantic masses
> of lumbering grey clouds overhead, the Chief Secretary (Mr.
> Chaffey) performed, on Friday, one of the most touching
> ceremonies of his career.[43]

Chaffey climbed the few steps of the hut, then turned and faced his small audience. He recounted the sad events that had led them here, spoke of the appropriateness of the memorial being an emergency shelter, and detailed some of the further steps being taken by the government to develop alpine infrastructure.[44] He then laid a wreath that had been sent by Laurie's family.[45] 'He Died Climbing', an inscription read.

Descending after the ceremonies, the party almost had their own disaster. The condition of the road had considerably worsened during the day. Ultimately one vehicle was forced to tow another back to the safety of the hotel, and a driver described making the trip back without functioning brakes.[46] The party was reportedly very glad to see the lights of the hotel. As they pulled their coats about themselves in the bitter wind and growing darkness, the attendees may have thought of their friends, lost in the cold and dark of a Kosciuszko storm. Because of early snows, and with the new telephone connections only due to be completed later in the year, Seaman's Hut was left isolated from the world for its first winter.[47]

There was one prominent figure associated with Laurie who did not attend the memorial's opening. On 8 January 1929, at a private hospital in Darlington in Sydney, she had given birth to 'a son'.[48] Nearly five months after his death, Laurie Seaman had become a father again.

Kindergarten Again

Chrissie named the boy Laurie, whereupon she slipped from the pages of the newspapers for a time. During the height of the search operations of August 1928, she had attracted some press attention, but the papers generally respected her privacy. Readers of the Adelaide News, for instance, knew at the time that:

> Under the terrible suspense, Mrs. W. L. Seaman, wife of one
> of the skiers ... is bearing up bravely ... She is in constant
> communication with the manager of the Hotel Kosciusko, and
> sits up practically the whole night waiting for telephone calls.[49]

Chrissie was generally peripheral to the intense focus on the search and rescue missions, and then ultimately the recovery and memorialisation of Laurie. For narrative purposes, she was the worried wife, and then the grieving widow, both of which were likely true, but which also simplified her experiences in service of her husband's story.

But she was not the only one to have her story simplified for emotional effect. When the chief secretary laid a wreath at the memorial hut in May 1929, he was reported saying 'that Laurie Seaman was not lost. He waited by the roadside for his companion'.[50] Although the exact circumstances between Laurie's summitting of Kosciuszko and death remained unclear, he was being held up as a beacon of mateship, finding death because he dutifully waited for a friend. It might have been true, but it was not demonstrably so.

As time went on the tragedy faded from the recorded activities of those concerned. Chrissie's own story again became clearer, somewhat independently of that of her late husband. Although not appearing in the social pages as regularly as she had at the time of her engagement and wedding a few years earlier, Chrissie was still active in the social life of her city. In February 1929 she remained an executive member of the local Sunbeam Kindergarten, serving as vice president.[51] In March she participated in a fundraising 'bridge afternoon' at a Vaucluse residence for the school.[52] Chrissie also attended other such bridge afternoons at Rose Bay in May, Bellevue Hill in June, and Alexandria in July.[53]

Only in passing connection did evidence of Chrissie's worries surface in the press. In April 1929 she donated five pounds and five shillings towards the Citizens' Southern Cross Rescue Fund.[54] This fund was organised for the search for Smith and Ulm, whose aeroplane the *Southern Cross* had seemingly disappeared in remote northern Australia. For a fortnight the crew of this plane were stranded and hungry. They were ultimately discovered alive, but while searching for the missing plane one of the rescue planes went down and two crew were killed. This too became subject to a search and rescue operation.[55] Like the case of Laurie and Evan at Kosciuszko, the story of the disappearance of the aviators in outback Australia attracted considerable press attention. The tragedy of the *Kookaburra* rescue plane became yet another of the stories that defined the epoch, feeding an image of the Australian continent as harsh and unforgiving. But amid that bigger story, Chrissie's gift points to her own experience of an angst-ridden rescue mission.

Despite this sign that she particularly empathised with events concerning the *Southern Cross* and *Kookaburra*, Chrissie's life continued to follow the routines of childrearing and charitable socialising. In December Chrissie was re-elected vice president of the Sunbeam Kindergarten committee.[56] But with the new year Laurie's story had another burst of interest, and Chrissie's husband was once again in the news. The Kosciuszko tragedy had one more act to run.

By Kosciuszko's Side

On 31 December 1929 shepherds who were watching sheep they had run into the high country for summer grazing were distracted

186

by the 'continual barking' of their dog.[57] A few hundred yards from Lake Cootapatamba, they found a body, about two miles from the spot where Laurie had died. The men notified the manager of the Kosciusko Hotel, who contacted the Jindabyne police, who then led a recovery party.

Sergeant Carroll headed this last search for Evan, soon described for the public by the *Sydney Morning Herald*:

> The scene of the tragic discovery was reached after a
> long climb. From the position of the body the police and
> experienced ski-ers were able to reconstruct, to some extent,
> the circumstances in which Mr. Hayes died. It is thought he
> lost his way in the blizzard, and became separated from his
> companion at Merritt's Lookout, to which place the tracks
> of the men were traced. He was apparently circling around
> in an endeavour to reach the summit of Mount Kosciusko to
> obtain his bearings when he was overcome by fatigue. His
> last act was to place his skis across each other to prevent
> himself from sinking in the snow when he lay down on them.
> One alpenstock was also found beneath the body, having
> apparently been used as a head rest, and the other was resting
> against a rock nearby.[58]

For sixteen months after the blizzard, Evan lay there through snowstorms and snowmelt, spring, summer, autumn, winter, and another spring and summer. When he was found, there was little to see. Giving evidence to the coronial inquest that followed, Carroll said that 'the body was in an advanced stage of decomposition; it was lying on its back, was fully clothed, the right arm was straight down parallel with the body, the left arm across the breast'.[59]

Evan was mostly skeletal, still wearing his tattered ski gear, and a gold signet ring 'on the third finger of the left hand'. This was inscribed 'W. E. H.' His skis too bore an inscription, 'E. H.' Evan's wallet contained water-damaged papers, a pair of goggles, and his 'Millions Club ski badge'.

After examining the body in situ, Carroll's party took it first to Betts Camp, where they spent the night, and, on 1 January 1930, to the Kosciusko Hotel. During the inquest, much attention was given to the question of whether any money had been found with Evan's remains. His former employer, the Aeolian Company, laid claim to '£50 in cash ... purported to be from the sale of furniture'. Nobody reported finding any such money. Only one of Evan's relatives was named in the proceedings, his brother Ronald, who 'resided at Wagga and was in the employ of Dalgety's Ltd'. After the inquest was concluded, Evan's last journey continued. He was taken down to Cooma and placed aboard the overnight mail train to Sydney.[60]

The next morning Evan was buried after a Requiem Mass at St Mary's Catholic Church in North Sydney.[61] Evan's brothers Gerald and Ronald attended, as did his sister Leah. Among the aunts and cousins mentioned by the newspapers, was a figure of particular note: 'Mrs. Laurie Seaman, of Palm Beach ... came especially to attend the service, to express her sympathy with the Hayes family, and to lay a wreath on the coffin.'

Many of Evan's skiing friends attended the service, and members of the Millions Ski Club acted as pallbearers. Among the congregation was Lorne Douglas, who had battled through the snow to send first news of the tragedy to the hotel.

After Mass, the coffin was carried from the church and conveyed to the cemetery. At the graveside, Fr James Magan S. J. spoke of the much-liked Evan. As with the larger tragedy itself, with different newspapers reporting separate parts of the address, the priest's tribute can be partially reconstructed:

> His aims were always high. ... As a boy he set great example
> in his service ... at the altar. He aimed at the highest
> spirituality – the Cross. On his last voyage he made what
> the world would call a great achievement, and he set a great
> example to all of us. There were peculiar circumstances of
> his life and death. He was born on Christmas day, and his
> remains were found, after many months, in the Christmas
> season[62] ...
>
> He was a man of rigid determination to do right. ...
> None aimed higher, and he reached the height of his mortal
> ambition when he conquered the snows of Kosciusko, to be
> in turn defeated by them, but I think he died where he would
> have liked to die – at the top.[63]

The much-loved skier was then lowered into the ground. Among the final gifts from friends and family, descending with the coffin was 'a floral ski'. About a week later a fellow student at Lithgow remembered Evan to the local press as 'a likable chap'.[64]

Several years later, in 1937, as the tenth anniversary of the tragedy neared, tenders were called for a memorial to Evan, but nothing on the scale of Seaman's Hut was ever built.[65]

In Memoriam

Like many stories from Australia's past, the tragic events of 1928 left the complex world of reality behind and became narrative history. Keith Bath, one of the would-be rescuers, penned a one-page account of the erection of Seaman's Hut and compiled short obituaries for Evan and Laurie, which were printed in the 1929 edition of *The Australian Ski Year Book*.[66] John Laidley, another of the men who searched for Evan and Laurie in August 1928, wrote a short account of 'The Kosciusko Fatality of 1928' for the same journal.[67] Bath's notes, with Laidley's ten-page article, became the basis for most subsequent recollections of what happened.

Although containing some errors, Laidley's short history remains a valuable source of information gleaned from participants, including Laidley himself. But it was penned before the discovery of Evan's body, so Laidley brought the story to its grim conclusion with a single page in the 1930 edition, admitting that he could offer little more explanation for what had happened to the two men. 'After all,' he wrote, 'we have had only an inch of evidence which we have been endeavouring to stretch into a yard of conjecture.'[68] 'What an end', he concluded, 'a man's death and a skier's death for both of them in the mountains they loved so much.'

For most of a decade the newspapers were quiet about the 1928 Kosciuszko tragedy, only occasionally mentioning it in passing. One 1933 report stands out for referring to a man rescued from his own tragedy by 'the Seaman–Hayes Memorial snow ambulance' – a sleigh pulled by four huskies.[69] But the story of the disappearance of Evan and Laurie generally appeared in simplified snippet form, serving as an occasional warning reference for skiing enthusiasts, or a quick backstory for the memorial hut that was itself a site of growing tourist interest. In fact, the moral of the story itself shifted with time, even as it was largely told in the same way.

With the ten-year anniversary, the 1928 Kosciuszko tragedy once again became a feature article. The *Sydney Mail* gave a page of its late August 1938 issue to recounting the events, admittedly largely drawn from Laidley's account.[70] But even while acknowledging that the details of the disappearance could only be supposed from slight evidence, the tragedy was held up 'as an example of two men's heroic endeavour to keep a mutual promise to meet at an appointed place, a promise which was to be a rendezvous with death'. In the absence of much detail, there was room for meaning to pile in. The story of disappearing mates had morphed into a parable about mateship.

This theme reappeared during 1941, when a June edition of *Smith's Weekly* again recounted the story.[71] By then Australia was fighting another war, which America had yet to join. In this context, the *Smith's Weekly* writer loaded the 1928 tragedy with significance drawn from the victims' international friendship. After telling the story, the writer referred to the memorial hut and concluded:

> It is a solid stone hut, red-roofed, a striking landmark
> visible for miles, a beacon to guide, a refuge from danger,
> and finally a symbol, a monument, to an Australian and an
> American who together faced the onrush of storm as to-day
> their respective nations face an onrush of the blizzard of
> world-war.[72]

Six months before Prime Minister John Curtin famously declared that 'Australia looks to America', the tragedy that befell Evan and Laurie was being held up as a symbol of friendship and alliance.[73]

The burning down of the Kosciusko Hotel in 1951 led to another round of reminiscence, albeit one still framed by repetition. In that year's 11 August edition of the *Sydney Morning Herald* Bartlett Adamson recounted 'A Tragic Mystery of Mount Kosciusko'.[74] In a few opening lines he mused on the lost hotel, the standing memorial hut, and got straight into the story as he knew it. Like his title, the article closely followed *Smith Weekly*'s 'The Mystery of Kosciusko' published a decade earlier. Besides keeping the general narrative structure, he replicated or closely paraphrased the text. 'Anxiety awakened', for instance, became 'Anxiety awoke'. But to be fair, Adamson was probably only plagiarising himself. He had been a regular writer for *Smith's Weekly*, and by repeating himself at least brought the story into another decade, albeit echoing Laidley's original theme of 'a mystery still unsolved'. The story was not merely an event, but a conundrum kept alive over the decades. In that, the Kosciuszko tragedy was history at its simplest, history as 'what happened' – both a question asked by the present and a sequence of events from the past.

Investigative historic treatment of Evan and Laurie's misadventure came towards the end of the twentieth century. Inspired by his first encounter with the memorial hut in 1976, journalist Robert Hefner wrote two articles about the tragedy for the *Canberra Times* in June 1982.[75] In the first, through using some of the historic newspaper reports, he added more detail to the story than the versions that tended to rely on Laidley's account. In the second, after visiting Glen Cove in America, he was able to flesh out some of Laurie's father's reaction to the tragedy. Similarly, in the early twenty-first century the memorial hut inspired continuing research into its originating story. Conservation architect David Scott critiqued Laidley's original account in a 26-page paper for the Kosciuszko Huts Association, completed in August 2013, further elaborating the details of the tragedy from newspapers and photographs.[76] With these and other retellings, the story could never seriously be considered forgotten. The memorial hut had fulfilled part of its key role.[77]

Yet the story of the tragedy also points out the caprice of historical memory, the amnesia of time. Despite the recollections and morality-rich commemoration – or because of them, perhaps – many aspects of the protagonists' lives rapidly drifted into the background or disappeared altogether. Laurie's meeting with Gandhi, the imprisonment of Evan's father, Chrissie's international courtship and other such life experiences remained beyond the scope of a story that tended to focus on a singular event and its lonely memorial: a tragedy at Kosciuszko on 14 August 1928, and a hut built in 1929.

But there was so much more to the story than that. A closer look at the people and place and events behind the erection of Seaman's Hut reminds us that the inscription is never the full story. In fact, it is often the start of a new chapter.

CHAPTER EIGHT

Myths and Realities

One way or another all Australians journey by Kosciuszko's sides. It pulls on our history as powerfully as the moon upon the tides. We all experience its effects, whether we go there or not.

When I made my own pilgrimage to the top of Kosciuszko at the start of 2018, my thoughts were of 1928. As I paused above Lake Cootapatamba I looked over the place where Evan had laid, and thought of his long rest by the mountain. From Seaman's Hut I gazed through a glass window at the rocky outcrops outside, signing the guestbook 'for Laurie' before I left.

I was probably alone on the mountain in thinking quite so deeply about these events and characters from our past. I saw walkers read the inscription at Seaman's Hut and the interpretative panels within. I noticed eyes glance at the old metallic camera under a glass case in the Thredbo Ski Museum, and saw people pause briefly before the dark monogrammed skis on the wall.

But these mementos told only a fraction of the story, prompting me to think of recent arguments about statues and memorials throughout our nation and that of Laurie's birth. I contemplated our fixations with famed explorers, shuddered at the persistence of a tired sort of national history that long-dead Professor Scott would recognise, and felt a little sorry for Evan and Laurie, who had captivated the country and then been mostly forgotten by it.

As I walked the ridges and valleys by and over the Snowy River I kept thinking of my own ancestral links. I wondered too if Evan had felt this pull of Kosciuszko, the strange summons of an inhospitable place. To walk towards Kosciuszko is, for me at least, to experience a sort of homecoming. Undoubtedly, I am not alone in that.

This is a place that has long pulled Australians to it. As I walked towards our mountain I thought of the scientists who had camped in wind and snow and sun to better understand our world. I recalled the trainloads of noisy children chugging up from Sydney and Brisbane. I contemplated the skiers and the walkers, the drivers and the riders. I thought of tribes and trackers. I mused over many stories that spoke of this place's special power. A game of cricket played at its base, a football trophy called the Kosciusko Cup.[1] An archbishop celebrating Sunday Mass on the summit, the congregation gathering with self-consciously biblical resonance.[2] Whether at play or at prayer, Kosciuszko touches our minds and hearts. We intuit its significance, even when we do not purposely contemplate what it might signify.

Eventually queuing for my own chance to be photographed by the summit cairn, I was struck by the polite camaraderie of the moment shared by so many strangers. People of all ages and races and genders and accents stood on Kosciuszko in a spirit of mutual purpose that would surely have satisfied the man for whom the mountain was named, pleased the ancestors who gathered for centuries near to it, and did so in a spirit in which Evan and Laurie would have shared.

And I knew then why I had been so drawn to this story and why it matters to us now.

Kosciuszko Dreams

Australia needs a new type of history willing to escape its own traditions. I felt that long before I started writing. My prior histories have all, in one way or another, tackled stories that we told imperfectly for a long time. We wrote national histories without ordinary people. We misdated our origins, misrepresented our deep past. We allowed division to grip us, separating our past into corners labelled academic or popular, left or right. But because history examines the past from the perspective of the present it can always be rewritten.

I started with that word which so conjures both place and people: Kosciuszko. Daydreaming, like clouds floating above the grassy plains of my youth, my mind drifted towards the mountains and was forced up. There, like mist condensing and freezing, the history took form and became a story of people and place. Having resolved the mountain as my focus, I soon found Evan and Laurie. Recognising elements of my world in theirs, I also saw theirs in mine. But it is the differences that are most instructive.

By getting lost and dying up by Kosciuszko, Evan and Laurie guide us into a world filled not with stale narratives and matey stereotypes but with nuance. Theirs was a history of Anzacs and tea ladies, Aboriginal trackers and Scandinavian accents, aeroplanes and trainloads of schoolchildren. Their world buzzed with firm whispers of the future and clear echoes of the colonial past. The nation's easy meta-myths of discovery, settlement and foreign bloodletting simply do not work in this landscape; they never did. Kosciuszko is a long way from the adventures of Captain Cook, and it is a better icon of our nation's history than any statue in any park in any town or city.

Fracturing our tendency to isolate and compartmentalise our own history, Evan and Laurie's friendship reminds us that Australian history is in fact deeply international. We are not some splendidly isolated exception at the bottom of the globe. The story of their going together to an Australian mountain named for a Polish hero of the American Revolutionary War speaks to a long association between Europe and Australia and the United States, enriched by their own personal histories. Laurie's travels and Evan's youth both reveal universal struggles of people and power, privilege and progress. These tensions were as evident in the factories of China as the coal mines of Lithgow.

Yet Kosciuszko also shows us that history and nations and people can be anchored by place. There is something distinctly Australian about Kosciuszko, more a feeling than an idea. This is the closest thing I can articulate to that meaning of Kosciuszko that I sensed at the top of the mountain. Like Sinai, Kosciuszko is a place where the threads of history and myth and spirit come together in a distinct community. I am sure that Evan and Laurie felt it too.

In their ski clubs and train rides and camping trips, Evan and Laurie and their fellow Australians shared the Kosciuszko experience. In their optimism for the future and sense of collective adventure, in those huge rescue efforts that followed their disappearance, Kosciuszko brought Australians together in body and mind and heart. In doing so, they created and fostered that very sense of togetherness for which Kosciuszko can well stand.

A well-travelled American Protestant born into ease and a Catholic Australian battler sought something at Kosciuszko that transcended the barriers of their worlds. Chrissie Seaman's letter of thanks to an Aboriginal tracker, the hurried rescue train that interrupted a working week, the aeroplanes fighting gales, the miners racing a blizzard, each story on Kosciuszko offered a peek into that thing often called the Australian spirit. We might also call it the spirit of Kosciuszko.

That spirit was in evidence before 1928, when soldiers took their photographs and songs of Kosciuszko to battle. It was in evidence when poets wrote of horsemen, and in those times both ancient and recent when moths were harvested for food and different words filled the alpine air. That spirit of Kosciuszko was there in cakes and parties, in fundraising and government-assisted holidays for children. That spirit filled the hotel, and spilled over onto the frozen pond and the snowy slopes, and in summer along the dusty track that cars took to the top of the Snowy River and beyond.

And that spirit survived the tragedy of 1928.

Kosciuszko Memories

Amid the crisscrossing human stories upon and beyond the mountain, those intimately connected with the 1928 tragedy offer a glimpse of Australian history seen through the lens of the mountain.

Continuing to be the ski expert of Kosciuszko, George Aalberg was frequently photographed showing his skills, especially by ski-jumping.[3] In 1930 he accompanied the governor-general on an excursion to Kosciuszko, although blizzard conditions meant they could not reach the summit.[4] Most of the party turned around at Seaman's Hut, while Aalberg took His Excellency up to the side of Kosciuszko proper before also turning back. When caught during a cross-country ski expedition in 1934, Aalberg survived another blizzard with two companions by building an ice shelter, then retreating into 'a hollow log' when the ice shelter failed.[5]

In 1935 he helped take a film crew to Kosciuszko, reportedly the first such expedition.[6] He worked for an indoor skiing facility at Rushcutters Bay in Sydney in 1936, allowing would-be skiers to learn some of the art before heading to Kosciuszko.[7] But after training and accompanying thousands of Australian skiers, the famous ski expert died in early 1939 aged only fifty-two.[8] Few had done more to popularise the sport in Australia than this carpenter from Norway, yet like Evan and Laurie before him, Aalberg quickly passed from the pages of the newspapers and popular memory.

Never as famous to begin with, at least in metropolitan circles, William Rutherford drew attention in the 1930s when he was dropped from the Dalgety police.[9] A hard-hitting article in the *Bombala Times*, ostensibly from *Truth*, argued that he was easy prey to the priority of budget 'because pennies must be saved – and he's black'. Among other services, his role in seeking Evan and Laurie was highlighted as proof of his professionalism, and Chrissie's letter of thanks was quoted as evidence of the sort of gratitude that it was felt he properly deserved. 'This man's dismissal has raised storms of protest everywhere', the journalist claimed, pointing out that pastoralists appreciated his employment as a disincentive to prospective stock thieves. Describing the scene, the writer dwelled on the injustice:

As the hour of five struck on the clock of the little stone
police station at Dalgety on the afternoon of December 5,
Rutherford, without a mark against his good name, walked
out into the world to look for another billet. So far as he was
concerned, the 15 years that he had spent aiding the police
with his keen eyes and native intelligence were wasted.[10]

Budgets were topical in the 1930s courtesy of the Great Depression.
A few months after Seaman's Hut was opened, the Wall Street Stock
Market crashed, precipitating global financial crisis. After visiting
America in 1932, Chrissie's father returned to Australia in early
1933 with stories of the trouble he had seen. As the ship bringing
him home stopped at various ports he was quizzed about America.
He told the press about empty office buildings that looked 'like a
tomb' and explained that 'farmers are nearly all bankrupt and the
banks are foreclosing on their properties in an endeavour to save
themselves from financial ruin.'[11] Bell explained that isolationist
and protectionist policies had evidently contributed to the cascading
crisis. 'America had failed to grasp the fact of the interdependence
of nations,' Bell argued to some fellow accountants, 'and her lone
hand policy would prove disastrous to herself'.[12]

Having attended a meeting of nations within the British Empire
at Ottawa to address the Great Depression, Bell was not some
uninformed observer. Through him, Laurie's family were again
close witnesses to the great events of history. Bell had gone to
America via England, in order to 'visit his 92-year-old mother'.[13]
With Chrissie and the boys in tow it was a unique opportunity to
connect four generations of this rather international family. But
it was very likely a bittersweet journey for Chrissie, because they

travelled by the *Ormonde*,[14] the ship on which she and Laurie had met and fallen in love.

Chrissie's return to Australia meant yet another return to the social pages. The Sydney *Sun* reported that she had been 'visiting her father-in-law, who built the chalet at Kosciusko in memory of his son Laurie'.[15] In April the *Sydney Mail* published her photograph, mentioning in a caption that she 'spends her time between her residence at Bellevue Hill and Palm Beach since her return'.[16] During May she was reportedly 'resting at Bowral for a week or two'.[17] She hosted a housewarming party at Point Piper in August at her new home in that suburb.[18] 'The color scheme throughout the reception rooms was in orange and yellow', and guests received memento cards and gifts. Chrissie 'gave one of the cheeriest cocktail parties of the week' in early October.[19] 'All the bright young things were there to fill the old house with music, song, and laughter,' said one participant.

At Palm Beach in December Chrissie 'introduced an American custom which typifies the spirit of Christmas', leading the way for what became an Australian tradition:

> A tree gaily decorated has been erected in the garden, and
> at sundown each evening the lights are switched on, with
> charming effect. The tree can be seen from the beach, the
> branches swaying in the breeze.[20]

Chrissie's Palm Beach house became something of an icon. Soon after winning praise for her illuminated Christmas tree, Chrissie's home was featured in a *Sun* piece on 'Harbour, Beach and Mountain Homes'.[21] Its centre page was a photograph of her house, 'which commands one of the most magnificent water views it is possible to imagine'. Complementing this were two photographs capturing the views, one of which had Chrissie seated on a verandah flanked by her two boys. An interior shot showed the billiard room of the house, balls on the table as if a game was left in progress. The overall image was of a modern building that took advantage of its situation, yet still had cosy spaces. Hinting at European comfort and alpine influences, it was generally known as 'the chalet'.

Forever linked to Kosciuszko by tragedy, Chrissie seems to have maintained some distance from the mountain. For instance, she did not participate in the festivities when the Millions Ski Club headed to the mountains in August 1934 for their annual and well-publicised holiday, even though the main competitive event was the 'W. Laurie Seaman Cup'.[22] Instead, she travelled to visit a friend in Colombo in Sri Lanka, crossing the Indian Ocean again aboard the memory-laden *Ormonde*.[23] For her, the ski season may have been too painful.

But the seasons rolled on and Australians still flocked to their mountain. Kosciuszko continued its pull upon Australia's imagination.

Our National Mountain

In the decades between these events and our own times, Kosciuszko quietly remained at the centre of Australian life and identity. One notable moment of reflection came in February 1940, when the centenary of Strzelecki's ascent and naming of Kosciuszko was marked.[24] The memorial was a simple bronze plaque, placed on granite upon the mountain's peak. The ceremony drew a large crowd of more than 120 schoolchildren, whose pennies and halfpennies had helped pay for the plaque, as well as the usual complement of politicians and dignitaries. Notable was the arrival of shepherds and cattlemen from the wider Snowy Mountains, 'hardy alpine men, riding their mountain brumbies' as one reporter put it.[25]

But in an event that could easily have been construed and constructed in triumphalist colonial terms, it was a foreigner who offered the clearest perspective on what the Australian mountain could mean. The plaque was unveiled by the consul-general of Poland, Ladislas Adam de Noskowski, and in his short speech he referred to the way that the mountain reminded Strzelecki of Tadeusz Kościuszko's tomb.[26] The plaque bore a transcription of Strzelecki's own explanation for the name. Prompted by its appearance, and recognising it was 'among free people who appreciated freedom', Strzelecki had said that 'I could not refrain from giving it the name of Mount Kosciusko'.[27]

The allusion to the tomb of Kościuszko had particular power at that time. Having led an uprising against Russian occupation in 1794, Kościuszko was an icon of liberty. Even in defeat he achieved a sort of poetic triumph, becoming a symbol of the good fight in a good cause (not unlike the way Gallipoli came to be seen by Australians), encapsulated in popular memory by the lines of the poet Thomas Campbell:

> Hope, for a season, bade the world farewell,
> And Freedom shriek'd – as KOSCIUSKO fell![28]

By 1940, when the Polish consul read Strzelecki's lines atop the mountain, Kościuszko had not only fallen, but his statues and monuments throughout Poland were being systematically destroyed by the Nazis.[29]

The Second World War again put Mount Kosciuszko in the minds of Australians overseas. Just as the slopes of Gallipoli had once been compared to Kosciuszko, so too was Kokoda.[30] Soldiers still used their mountain to measure the world and referred to Kosciuszko when they longed for home.[31] Kosciuszko again came to the aid of comfort fund activities, as when a table was decorated like the mountain and porcelain paratroopers with handkerchief parachutes were dropped nearby.[32] It served to represent international relations, as when a piece of rock from the summit was gifted to the American–Polish Committee in New York.[33] And, as already seen, Evan and Laurie's misadventure played its part in international alliance-making too.

Yet Kosciuszko also came to stand for a future peace. In 1943 the government of New South Wales announced plans to create a 1,140,000-acre state park.[34] Explaining this proposal, Premier William J. McKell wrote:

> it is a first consideration that our important river resources
> should be protected from deterioration and pollution, and that
> our park and recreation lands should be conserved for the
> use of the people. ... The Major function of a park area is the
> promotion of the fitness and well-being of Australians through
> the health-giving qualities of relaxation and recreation in a
> natural atmosphere of inspiring grandeur.[35]

Expressed in these ideals, and enacted with the *Kosciusko State Park Act* of 1944, Kosciuszko thus came to focus a vision of postwar Australia.[36]

These ideals were not realised without problems. While the new chalet built at Charlotte Pass in 1939 was hailed for its architectural novelty and touristic convenience, it was also a site for industrial dispute and forebodingly notable for its use of 'Asbestos

Cement Sheeting'.[37] The ski fields were further developed, giving more Australians their chance to access and experience the 'big Alps stuff', but also raised questions about protection and access and what exactly constitutes Kosciuszko's 'natural atmosphere'. The sheep and cattle are no longer grazed there. The bogong moths remain largely uneaten. Even the wild brumbies face periodic threats of a cull.

Talk of Kosciuszko's 'natural atmosphere' was, unsurprisingly, also linked with political doublespeak. Through that, it now stands for a truth that should transcend politics. In 1944 one writer for the *Sydney Morning Herald* pointed out that the government's state park scheme was not simply about tourism. 'Primarily,' he said,

> this remote region was placed under the control of a national
> trust to protect the vast watersheds that feed the Snowy River,
> so ensuring the continuity of the water supply which would be
> the pivot of two rival projects, the Snowy hyrdo-electric and
> Snowy irrigation diversion schemes.[38]

That those schemes have brought the nation economic benefits, environmental degradation, industrial success and ongoing political headaches, show some of the many ways that Kosciuszko has and continues to play an important part in our national story in the most structural sense. In fact, much of this has its origins in the 1920s, when the Kosciusko Hotel was first powered by a small hydro-electrical generator.[39]

Yet there is a more important part to this industrial side of greater Kosciuszko: that of identity. Through the Snowy Mountains hydroelectric works, and the irrigation schemes they supported in the plains out west, the very face of Australia was changed. The manpower it required helped accelerate the end of the White Australia Policy, the mountain experiences of such immigrants helped make Kosciuszko their own. In doing so, they made its continent their home, much as my distant ancestors had done.

In 1967 Kosciuszko National Park was established as part of the creation of the new National Parks and Wildlife Service.[40] The park's establishment in that year serendipitously links the mountain with an important change in Australia's attitude to and treatment of its First Peoples. The journey to Kosciuszko is now marked by interpretative panels that explain to visitors the Aboriginal history and significance of this remarkable landscape.

Kosciuszko now serves as it has always done, as the highest vantage point from which to survey Australia's past and present, and perhaps also its future. In the nineteenth century, Strzelecki described the summit of Kosciuszko:

> the snowy and craggy sienitic cone of Mount Kosciuszko
> is seen cresting the Australian Alps, in all the sublimity of
> mountain scenery. Its altitude reaches 6500 feet, and the view
> from its summit sweeps over 7000 square miles.[41]

As he turned at the top of Kosciuszko, he remarked upon the country even beyond his sight. East was the rich coast, west ran mighty rivers. For him, Kosciuszko was worth naming for an icon of freedom because it looked over a land of promise.

Over a century later, Elyne Mitchell stood on one of the adjacent peaks, and with Kosciuszko before her mused on those emotions that such places inspire within us:

What is it that is so utterly satisfying about having gained
a mountain peak – from that well-earned eyrie gazing out
over miles of hills and ridges, feeling how far removed one
is from the peace of the sun-filled valleys, appreciating their
peacefulness and yet longing to hold to oneself for ever
something of the curbed strength of the wilderness, trying to
engrave for ever on one's mind the lift and fold of the sun-
bathed ridges, the mountains quiescent that will soon be white
with storm?[42]

She understood that pull of the mountain, that urge to embrace a
discomforting sort of comfort, that sense that time cannot stand
still, that search for clarity.

A generation or so later I had my first experience of Kosciuszko.
It remains one of my clearest childhood memories. I was standing at
Charlotte Pass with my family, looking into the wilderness beyond
the road. Straining, I hoped at least to glimpse our highest mountain
far beyond the carpark. I argued to go closer. But I was too young
and the weather too ominous, so I was bundled back into the car,
and we turned away. But I still remember the dull white snow and
the darkening skies, and that desperate longing for Kosciuszko.

Endnotes

Preface

1   *Sydney Mail*, 31 August 1927, p. 13.

2   *Sydney Morning Herald*, 29 January 1852, p. 2.

3   T. W. Mitchell, 'Baal Udthu Yamble Yabba', *The Australian Ski Year Book*, 1953, pp. 66–69.

4   Ernest Scott, *A Short History of Australia*, 4th edition (Melbourne: Humphrey Milford, 1920), p. 321.

5   Scott, *A Short History of Australia*, pp. 76, 229.

6   *Canberra Times*, 11 January 1939, p. 4.

7   Scott, *A Short History of Australia*, p. v.

Chapter One   A Kosciuszko Childhood

1   *Queanbeyan Age*, 29 February 1884, p. 2; see also my *Kin*, pp. 174–79.

2   *Albury Banner and Wodonga Express*, 23 August 1907, p. 12.

3   *Albury Banner and Wodonga Express*, 20 September 1907, p. 13.

4   *Albury Banner and Wodonga Express*, 24 July 1908, p. 12.

5   *Albury Banner and Wodonga Express*, 30 October 1908, p. 13.

6   *The Australian Ski Year Book*, 1931, pp. 53–56.

7   *Australian Town and Country Journal*, 19 June 1907, p. 31.

8   *Australian Town and Country Journal*, 5 September 1906, p. 33.

9   *Albury Banner and Wodonga Express*, 28 February 1908, p. 13.

10 *Albury Banner and Wodonga Express*, 9 October 1908, p. 12.

11 *Australian Town and Country Journal*, 28 November 1906, p. 33.

12   *Australian Town and Country Journal*, 5 September 1906, p. 33.

13   *Sydney Morning Herald*, 20 February 1906, p. 5.

14   *Daily Telegraph*, 17 January 1907, p. 2; *Shoalhaven News and South Coast District Advertiser*, 23 March 1907, p. 2.

15   *Sydney Morning Herald*, 23 February 1907, p. 12.

16   *Tocumwal Guardian and Finley Free Press*, 8 January 1909, p. 5; *Sydney Morning Herald*, 9 June 1909, p. 8.

17   *Sydney Morning Herald*, 9 June 1909, p. 8.

18   *Albury Banner and Wodonga Express*, 27 March 1908, p. 13.

19   *Manaro Mercury, and Cooma and Bombala Advertiser*, 23 March 1908, p. 2.

20   *Sydney Morning Herald*, 30 July 1913, p. 13; *Daily Telegraph*, 30 July 1913, p. 6; *Northern Star*, 4 August 1913, p. 6. See also: *Sydney Morning Herald*, 30 December 1914, p. 7; *Queanbeyan Age*, 28 July 1915, p. 6.

21   *Sydney Morning Herald*, 30 July 1913, p. 13.

22   Ibid.

23   New South Wales Registry of Births, Deaths and Marriages, 2633/1907, 26584/1909.

24   *Sydney Morning Herald*, 18 April 1911, p. 5.

25   *Sydney Morning Herald*, 16 January 1911, p. 9.

26   *Sydney Morning Herald*, 25 January 1911, p. 8.

27   *Sydney Morning Herald*, 22 April 1911, p. 18.

28   *Lithgow Mercury*, 8 July 1912, p. 2.

29   *Daily Telegraph*, 10 February 1911, p. 8.

30   *Lithgow Mercury*, 10 February 1911, p. 4.

31   *Daily Telegraph*, 16 February 1911, p. 9.

32   *Lithgow Mercury*, 4 August 1911, p. 4.

33   Ibid.

34   *Lithgow Mercury*, 30 August 1911, p. 2.

35   Ibid.

36   *Lithgow Mercury*, 13 September 1911, p. 2.

37   *Lithgow Mercury*, 30 August 1911, p. 2.

38   *Lithgow Mercury*, 6 September 1911, p. 2.

39   *Lithgow Mercury*, 13 September 1911, p. 2.

40   *Lithgow Mercury*, 15 September 1911, p. 4.

41   *Lithgow Mercury*, 16 October 1911, p. 2.

42   *Lithgow Mercury*, 13 October 1911, p. 4.

43   *Cumberland Argus and Fruitgrowers Advocate*, 8 June 1901, p. 3.

44   *Cumberland Argus and Fruitgrowers Advocate*, 22 March 1899, p. 2.

45   *Bathurst Times*, 18 October 1911, p. 2.

46 *Lithgow Mercury*, 1 November 1911, p. 2.

47 Ibid; 14 February 1912, p. 2; 13 March 1912, p. 1.

48 *Lithgow Mercury*, 3 June 1912, p. 2.

49 *Lithgow Mercury*, 7 June 1912, p. 4.

50 *Lithgow Mercury*, 5 July 1912, p. 4.

51 *Lithgow Mercury*, 8 July 1912, p. 2.

52 *The Australian Ski Year Book*, 1929, pp. 156–57.

53 *Sydney Morning Herald*, 31 August 1921, p. 8.

54 *Sydney Morning Herald*, 13 December 1921, p. 1.

55 Ibid.

56 *Sydney Morning Herald*, 16 December 1925, p. 13.

57 *Sydney Morning Herald*, 26 June 1925, p. 8.

58 *Sydney Morning Herald*, 29 November 1927, p. 5.

59 *Sydney Morning Herald*, 4 October 1927, p. 6.

60 *Evening News*, 10 December 1920, p. 1.

61 *Sydney Morning Herald*, 31 August 1921, p. 8; *Daily Telegraph*, 31 August 1921, p. 4.

62 *Evening News*, 5 May 1922, p. 8.

63 *Sunday Times*, 11 February 1923, p. 15.

64 New South Wales Registry of Births, Deaths and Marriages, 3466/1923; *Sydney Morning Herald*, 24 February 1923, p. 14.

65 *Evening News*, 7 May 1923, p. 11; *Sunday Times*, 8 July 1923, p. 27.

66 *Sunday Times*, 7 October 1923, p. 18.

67 *Newcastle Morning Herald and Miners' Advocate*, 1 February 1924, p. 10; *Sunday Times*, 24 February 1924, p. 7; *Sun*, 18 March 1924, p. 10; *Sunday Times*, 23 March 1924, p. 14; *Evening News*, 1 December 1924, p. 13.

68 *Sun*, 26 March 1925; *Truth*, 14 November 1926, p. 11.

69 *Sun*, 15 May 1927, p. 37.

70 *The Australian Ski Year Book*, 1929, pp. 156–57.

71 *Sun*, 8 August 1920, p. 14.

72 *Sun*, 6 August 1922, p. 5; *Sun*, 7 August 1922, p. 7.

73 *The Australian Ski Year Book*, 1928, p. 1; *Sydney Morning Herald*, 21 July 1928, p. 10.

74 *The Australian Ski Year Book*, 1928, pp. 11–12. The contents page gives Evan as the author, and it is mostly likely that the 'popular secretary' is a reference to him, as he was secretary of the Millions Ski Club. S. A. Bailey was one of the secretaries of the Millions Club, which may have been the cause of some editorial confusion.

75 *The Australian Ski Year Book*, 1929, pp. 156–57.

76   *Tumut Advocate and Farmers and Settlers' Adviser*, 26 August 1924, p. 2.
77   *Sydney Morning Herald*, 19 August 1926, p. 14.
78   *The Australian Ski Year Book*, 1928, pp. 11–12.
79   *Queensland Times*, 6 September 1926, p. 7.
80   Ibid.
81   *Evening News*, 26 August 1925, p. 16.
82   *Sun*, 6 August 1922, p. 5.
83   Ibid.
84   *The Australian Ski Year Book*, 1929, pp. 156–57.
85   *Evening News*, 22 August 1930, p. 6.
86   *The Australian Ski Year Book*, 1929, pp. 156–57.
87   *The Australian Ski Year Book*, 1928, pp. 43–47.
88   *Sydney Morning Herald*, 7 September 1897, p. 3.
89   Ibid.
90   *Sunday Times*, 19 August 1928, p. 1.

Chapter Two   American Liberty
1   *The Halcyon of 1916* (Swarthmore, 1916), pp. 41, 44, 148.
2   Ibid, pp. 54, 73, 89, 95, 96, 200.
3   *Transactions of the Institution of Engineers, Australia*, vol. 9, 1928, pp. 277–78.
4   *Brooklyn Daily Eagle*, 13 December 1917, p. 15.
5   *Brooklyn Daily Eagle*, 10 December 1919, p. 8; 22 March 1921, p. 7.
6   *The Australian Ski Year Book*, 1929, pp. 157–58.
7   *Brooklyn Daily Eagle*, 28 January 1920, p. 10.
8   *Transactions of the Institution of Engineers, Australia*, vol. 9, 1928, pp. 277–78.
9   *The Australian Ski Year Book*, 1929, pp. 157–58.
10   *Chanute Daily Tribune*, 22 January 1923, p. 1.
11   *Topeka State Journal*, 13 July 1922, p. 10; *Brooklyn Daily Eagle*, 5 October 1923, p. 9.
12   In January 1923, for instance, the *Pittsburgh Daily Post* advertised a series of forthcoming articles by Drew on 'What the Orient means to the American', *Pittsburgh Daily Post*, 21 January 1923, p. 10.
13   *Chanute Daily Tribune*, 22 January 1923, p. 3.
14   Ibid.
15   Ibid.
16   *Delaware County Daily Times*, 24 February 1923, p. 9.
17   *Friends' Intelligencer*, 24 March 1923, pp. 203–04.
18   *Friends' Intelligencer*, 24 March 1923, p. 203.

19   *Pittsburgh Daily Post*, 28 January 1923, p. 52; *Chanute Daily Tribune*, 31 January 1923, p. 3; *Pittsburgh Daily Post*, 11 February 1923, p. 20; *Chanute Daily Tribune*, 12 February 1923, p. 3; *Pittsburgh Daily Post*, 17 June 1923, p. 66; 15 July 1923, p. 18.

20   *Pittsburgh Daily Post*, 15 July 1923, p. 18.

21   *Brooklyn Daily Eagle*, 17 December 1922, p. 4E.

22   Ibid.

23   Ibid.

24   Ibid.

25   Ibid.

26   Ibid.

27   Ibid.

28   *Wilkes-Barre Times Leader, the Evening News*, 20 April 1923, p. 16.

29   Ibid., p. 2.

30   *Pittsburgh Daily Post*, 29 July 1923, p. 50.

31   *Wilkes-Barre Times Leader, the Evening News*, 30 April 1923, p. 5.

32   Ibid.

33   *Auckland Star*, 31 January 1923, p. 5.

34   *Wilkes-Barre Times Leader, the Evening News*, 11 June 1923, p. 11.

35   Ibid.

36   *New Zealand Herald*, 10 February 1923, p. 1 (Supplement).

37   Ibid.

38   *Wilkes-Barre Times Leader, the Evening News*, 11 June 1923, p. 11.

39   *Pittsburgh Post*, 12 August 1923, p. 50.

40   Ibid.

41   See for instance: *Temuka Leader*, 20 February 1923, p. 3; *Temuka Leader*, 22 February 1923, p. 3.

42   *Gippsland Times*, 9 April 1923, p. 2.

43   *Geelong Advertiser*, 12 May 1923, p. 6; *Mercury*, 28 May 1923, p. 3; *Examiner*, 29 May 1923, p. 3.

44   *Riverine Herald*, 15 May 1923, p. 2.

45   *Pittsburgh Daily Post*, 18 November 1923, p. 20.

46   Ibid.

47   Ibid.

48   *Pittsburgh Daily Post*, 26 August 1923, p. 17.

49   Ibid.

50   *Sydney Morning Herald*, 9 March 1923, p. 10.

51   Ibid.; 'Marjorie' seems to have been another nickname or a newspaper misidentification, as this was clearly Joyce Russell.

52   *Sydney Morning Herald*, 8 February 1923, p. 5.

53   *Sydney Morning Herald*, 12 February 1923, p. 4.

54   Ibid.

55   *Sydney Morning Herald*, 10 March 1923, p. 8.

56   *Daily Telegraph*, 16 January 1917, p. 3.

57   *Sydney Morning Herald*, 12 January 1917, p. 8.

58   National Archives of Australia, B2455, BELL R P [3009046].

59   State Library of New South Wales, MLMSS 886.

60   State Library of New South Wales, MLMSS 886, pp. 36–37.

61   *Sunday Times*, 18 February 1917, p. 28.

62   *Freeman's Journal*, 19 April 1917, p. 29.

63   *Sunday Times*, 24 February 1918, p. 13.

64   *Manning River Times and Advocate*, 20 August 1913, p. 3.

65   *Sunday Times*, 1 September 1912, p. 4.

66   *Daily Telegraph*, 14 April 1919, p. 3.

67   *Sydney Morning Herald*, 15 June 1922, p. 5.

68   *Evening News*, 9 March 1923, p. 11.

69   *Daily Commercial News and Shipping List*, 12 March 1923, p. 8.

70   *Argus*, 19 March 1923, p. 10.

71   *Daily News* (Perth), 23 March 1923, p. 8.

72   *Sunday Times*, 13 May 1923, p. 15.

73   *Daily Commercial News and Shipping List*, 3 April 1923, p. 8; 13 April 1923, p. 8.

74   *Brooklyn Daily Eagle*, 5 October 1923, p. 9.

75   *Daily Commercial News and Shipping List*, 13 April 1923, p. 8.

76   *Sun*, 6 January 1923, p. 23.

77   *Daily Commercial News and Shipping List*, 19 April 1923, p. 8; 20 April 1923, p. 8; 23 April 1923, p. 8; 26 April 1923, p. 8; 27 April 1923, p. 8.

78   National Archives (UK), *Board of Trade: Commercial and Statistical Department and Successors: Inwards Passenger Lists,* Class: BT26; Piece: 742; Item: 11.

79   *Pittsburgh Daily Post*, 21 October 1923, p. 16; 4 November 1923, p. 15; see also *Pittsburgh Daily Post*, 20 January 1924, p. 14.

80   National Archives (UK), *Board of Trade: Commercial and Statistical Department and Successors: Inwards Passenger Lists,* Class: BT26; Piece: 742; Item: 75; *Daily Commercial News and Shipping List*, 24 April 1923, p. 8.

81   *Brooklyn Daily Eagle*, 21 August 1923, p. 17.

82   *Brooklyn Daily Eagle*, 5 October 1923, p. 9.

83   *Brooklyn Daily Eagle*, 21 August 1923, p. 17.

84   *Pittsburgh Daily Post*, 25 November 1923, p. 20.

ENDNOTES

Chapter Three   The Millions Club

1   *Truth* (Brisbane), 1 February 1925, p. 11.
2   A typographical error in the original is corrected here.
3   *Sunday Times* (Perth), 6 July 1924, p. 32; *Advocate* (Burnie), 23 October 1926, p. 16; *Morning Bulletin* (Rockhampton), 8 January 1927, p. 14.
4   *Sun*, 13 August 1931, p. 15.
5   *Australian Woman's Mirror*, 22 February 1927, p. 44; *Advertiser and Register* (Adelaide), 14 April 1931, p. 11.
6   *Daily Examiner*, 19 November 1920, p. 1; *Sydney Morning Herald*, 2 February 1920, p. 13; *Labor Daily*, 7 July 1925, p. 6.
7   *Sydney Morning Herald*, 25 October 1921, p. 6.
8   *West Australian*, 7 March 1924, p. 1.
9   A. B. Paterson, *The Works of 'Banjo' Paterson* (London: Wordsworth, 2008), p. 6.
10   Nick Brodie, *Kin: A Real People's History of Our Nation* (Melbourne: Hardie Grant, 2015), pp. 98–101.
11   *Picton Post*, 7 September, 1933, p. 3; *Forbes Advocate*, 19 April 1940, p. 7.
12   *The Works of 'Banjo' Paterson*, pp. 32–33, 46.
13   Ibid, p. 3.
14   Brodie, *Kin*, p. 95–97.
15   *The Works of 'Banjo' Paterson*, pp. 226–28.
16   Ibid., p. 235.
17   'Nothing Unusual: The Weather and the Prophets', *Herald*, 8 June 1926, p. 6.
18   'The Winds', *Sydney Mail*, 23 May 1923, p. 23.
19   For example: *Daily Telegraph*, 26 January 1926, p. 5; *Sydney Morning Herald*, 15 November 1926, p. 11; *Daily Telegraph*, 1 December 1928, p. 19.
20   F. W. Tidd, *Dawn on Kosciusko Waltz* (Sydney and Bega: Braitling & Co., 1907).
21   Stefan Polotynski, *Kosciusko: Waltz for Pianoforte* (Sydney: W. J. Deane & Son, 1921).
22   *Armidale Chronicle*, 16 November 1921, p. 4.
23   *Brooklyn Daily Eagle*, 30 November 1923, p. 7.
24   *Sunday Times*, 9 December 1923, p. 24.
25   *Sunday Times*, 27 January 1924, p. 18.
26   *Sunday Times*, 13 July 1924, p. 17.
27   *Sunday Times*, 6 July 1924, p. 18; 13 July 1924, p. 18.
28   *Daily Telegraph*, 17 July 1924, p. 9.
29   *Evening News*, 1 May 1925, p. 5.

215

30   See e.g. *Grenfell Record and Lachlan District Advertiser*, 4 April 1913, p. 2; *Daily Telegraph*, 19 September 1913, p. 15; *Murrumbidgee Irrigator*, 27 April 1917, p. 3; *Armidale Express and New England General Advertiser*, 15 July 1919, p. 4; *Maitland Mercury*, 9 July 1920, p. 3.

31   *Transactions of the Institution of Engineers, Australia*, vol. 9, 1928, pp. 277–78.

32   *Sydney Morning Herald*, 6 November 1926, p. 27; By late autumn 1928 they also had an office in the Riverina at Narrandera, suggestive of considerable success: *Narandera [sic] Argus and Riverina Advertiser*, 22 May 1928, p. 2.

33   *Evening News*, 10 September 1925, p. 19.

34   *Sun*, 4 June 1925, p. 15.

35   *Sun*, 2 December 1926, p. 21; *Sydney Morning Herald*, 2 December 1926, p. 5; *Evening News*, 2 December 1926, p. 23.

36   *Evening News*, 30 June 1926, p. 10.

37   *Sun*, 30 June 1926, p. 8.

38   *Evening News*, 26 June 1926, p. 14.

39   *Sun*, 7 August 1927, p. 28.

40   Northern Beaches Library, Pittwater Image Library, Files PB/PB141-182.

41   *Sydney Morning Herald*, 9 April 1927, p. 10; *Evening News*, 9 April 1927, p. 8.

42   *Evening News*, 23 February 1927, p. 18; *Sydney Morning Herald*, 2 March 1927, p. 9; *Sun*, 2 March 1927, p. 17; *Evening News*, 8 March 1927, p. 12; *Sydney Morning Herald*, 31 March 1927, p. 12; *Sun*, 22 May 1927, p. 26; *Sydney Morning Herald*, 25 May 1927, p. 8; *Sun*, 27 May 1927, p. 3; 7 August 1927, p. 28; *Sunday Times*, 30 October 1927, p. 15.

43   *Sun*, 12 August 1926, p. 12; *Sydney Morning Herald*, 13 August 1926, p. 10.

44   *Mail*, 14 August 1926, p. 5; *News*, 14 August 1926, p. 4; *Advertiser*, 16 August 1926, p. 13; *Register*, 16 August 1926, p. 8.

45   *Register*, 11 August 1926, p. 10.

46   *Country Life Stock and Station Journal*, 14 March 1924, p. 3; *Sun*, 28 February 1928, p. 10; *Tumut and Adelong Times*, 24 January 1928, p. 6.

47   *Sydney Morning Herald*, 19 October 1927, p. 18.

48   *The Australian Ski Year Book*, 1929, pp. 157–58.

49   *Sydney Morning Herald*, 12 July 1928, p. 5.

50   *Sydney Morning Herald*, 19 October 1927, p. 18.

51   *Sydney Morning Herald*, 11 December 1894, p. 6; *Coolgardie Miner*, 11 November 1897, p. 6.

52 *Sunday Times*, 31 March 1907, p. 5.

53 Ibid.

54 Ibid.

55 *Sydney Morning Herald*, 28 November 1911, p. 9.

56 *Sydney Morning Herald*, 6 January 1912, p. 12.

57 *Sydney Morning Herald*, 24 June 1926, p. 12.

58 *Sydney Morning Herald*, 20 June 1928, p. 16, 18.

59 *Sydney Morning Herald*, 19 June 1928, p. 10.

60 *Age*, 10 August 1928, p. 10.

61 Ibid.

62 *Sydney Morning Herald*, 15 March 1928, p. 6.

63 Ibid.

64 *Sydney Morning Herald*, 19 March 1928, p. 16; *Referee*, 21 March 1928, p. 17.

65 *Sydney Morning Herald*, 23 April 1928, p. 8.

66 *Truth*, 22 April 1928, p. 7; *Sunday Times*, 22 April 1928, p. 40; *Sun*, 22 April 1928, p. 14.

67 *Sun*, 12 May 1928, p. 8.

68 *Sunday Times*, 13 May 1928, p. 18; *Sydney Morning Herald*, 14 May 1928, p. 6; 15 May 1928, p. 6; 17 May 1928, p. 6.

69 *Sydney Morning Herald*, 14 May 1928, p. 6.

70 *Sydney Morning Herald*, 17 May 1928, p. 6.

71 *Sydney Morning Herald*, 14 May 1928, p. 6.

72 *Sunday Times*, 13 May 1928, p. 18.

73 *Sydney Morning Herald*, 16 May 1928, p. 19.

74 *Sydney Morning Herald*, 15 May 1928, p. 6.

75 *Sydney Morning Herald*, 11 July 1928, p. 20.

76 *Sydney Morning Herald*, 12 July 1928, p. 8; 14 August 1928, p. 7.

77 *Sunday Times*, 5 August 1928, p. 26.

78 *Sun*, 10 August 1928, p. 3.

79 *Sun*, 9 August 1928, p. 15.

80 *Sydney Morning Herald*, 30 May 1889, p. 4; *Sydney Morning Herald*, 31 May 1889, p. 8.

81 *Manaro Mercury*, 30 March 1900, p. 2; *Sydney Morning Herald*, 8 February 1902, p. 7; *Sydney Morning Herald*, 5 August 1910, p. 3; *Queanbeyan Observer*, 23 April 1912, p. 2.

82 *Sunday Times*, 16 August 1925, p. 4.

83 *Sydney Morning Herald*, 13 August 1926, p. 8; *Evening News*, 12 August 1926, p. 4; *Sun*, 6 August 1927, p. 7.

84 *Sydney Morning Herald*, 24 July 1926, p. 11.

85   *Gosford Times and Wyong District Advocate*, 26 July 1928, p. 2.
86   *Goulburn Evening Penny Post*, 15 December 1925, p. 2; *Sydney Morning Herald*, 16 December 1925, p. 15; see also *Sun*, 10 December 1925, p. 3; *Goulburn Evening Penny Post*, 17 December 1925, p. 2.
87   *Goulburn Evening Penny Post*, 10 August 1928, p. 2.
88   *Don Dorrigo Gazette and Guy Fawkes Advocate*, 20 July 1928, p. 6.
89   *Gosford Times and Wyong District Advocate*, 26 July 1928, p. 2.
90   *Warwick Daily*, 6 July 1928, p. 3.
91   *Gosford Times and Wyong District Advocate*, 26 July 1928, p. 2.
92   *Sun*, 26 August 1928, p. 2.
93   *Gosford Times and Wyong District Advocate*, 26 July 1928, p. 2.
94   *Sydney Mail*, 18 April 1928, pp. 63–64.
95   *Sun*, 26 August 1928, p. 2.
96   *Sydney Morning Herald*, 24 July 1926, p. 11.
97   *Sun*, 26 August 1928, p. 2.
98   *Daily Telegraph*, 14 October 1924, p. 2; *Labor Daily*, 15 February 1926, p. 1.
99   *Brisbane Courier*, 11 December 1925, p. 13; *Register*, 5 January 1926, p. 5; *Evening News*, 30 January 1926, p. 8; *News*, 1 November 1928, p. 8.
100 *Saturday Journal*, 10 July 1926, p. 13.
101 Ibid.
102 Ibid.
103 *Canberra Times*, 4 January 1930, p. 2.
104 *Don Dorrigo Gazette and Guy Fawkes Advocate*, 20 July 1928, p. 6.
105 *Armidale Express and New England General Advertiser*, 27 July 1928, p. 5.
106 Ibid.
107 *Don Dorrigo Gazette and Guy Fawkes Advocate*, 20 July 1928, p. 6.
108 *Warwick Daily News*, 6 July 1928, p. 3.
109 *Don Dorrigo Gazette and Guy Fawkes Advocate*, 20 July 1928, p. 6.
110 *Armidale Express and New England General Advertiser*, 27 July 1928, p. 5.
111 *Warwick Daily News*, 6 July 1928, p. 3.
112 *Sunday Times*, 19 August 1928, p. 1.

Chapter Four   Stormclouds
1   *Sydney Morning Herald*, 9 January 1922, p. 8.
2   Ibid.
3   *Education: Journal of the N.S.W. Public School Teachers Federation* 6(7), 15 May 1925, p. 186.

4   *The Bulletin* 22(1129), 5 October 1901, p. 18.

5   *Sydney Morning Herald*, 12 February 1897, p. 4; *Sydney Morning Herald*, 17 December 1897, p. 6; *Sydney Morning Herald*, 18 December 1897, p. 8.

6   *Evening News*, 14 May 1898, p. 4; *Evening News*, 19 September 1989, p. 7.

7   *Daily Telegraph*, 29 January 1907, p. 7.

8   *Sydney Morning Herald*, 2 October 1900, p. 7; *Sydney Morning Herald*, 27 March 1903, p. 6; *Sydney Morning Herald*, 16 May 1906, p. 6.

9   *The Methodist*, 23 June 1906, p. 2.

10 *Evening News*, 23 January 1907, p. 2.

11 *Sydney Mail*, 11 July 1928, p. 27; *Armidale Express and New England General Advertiser*, 27 July 1928, p. 5.

12 *Sunday Times*, 25 July 1926, p. 5.

13 *Dubbo Dispatch and Wellington Independent*, 6 December 1921, p. 2; National Archives of Australia, B2455, DOUGLAS L. K. [3517277].

14 C. Larsen, in fact, but his identity remains elusive. Subsequent reports affirm fellow skier Emil Sodersteen's identification of Larsen over other conflicting reports. See: *Evening News*, 17 August 1928, p. 1.

15 *Sydney Mail*, 8 August 1928, p. 12.

16 *Daily Telegraph*, 19 August 1905, p. 5; *Sydney Morning Herald*, 24 September 1913, p. 9.

17 Poppy Biazos Becerra and Peter Reynolds, 'Sodersten, Emil Lawrence (1899–1961)', Australian Dictionary of Biography, National Centre of Biography, Australian National University, http://adb.anu.edu.au/biography/sodersten-emil-lawrence-11734/text20979.

18 *Construction and Local Government Journal*, 15 June 1927 p. 12; *Construction and Local Government Journal*, 28 March 1928, p. 15.

19 *Sydney Morning Herald*, 18 August 1928, p. 17 (Sodersteen's account); 17 August 1928, p. 11; *The Australian Ski Year Book*, 1929, pp. 97–8.

20 *Sydney Morning Herald*, 17 August 1928, p. 11; *The Australian Ski Year Book*, 1929, pp. 97–98.

21 *Sydney Morning Herald*, 18 August 1928, p. 17.

22 *Sun*, 10 August 1928, p. 3.

23 *Sun*, 1 August 1928, p. 8.

24 *Argus*, 8 August 1928, p. 23.

25 *Sun*, 8 August 1928, p. 8.

26 *Argus*, 9 August 1928, p. 15; 10 August 1928, p. 15; 11 August 1928, p. 15; 13 August 1928, p. 15; 14 August 1928, p. 15; 15 August 1928, p. 21.

27 *Argus*, 15 August 1928, p. 20.

28 Ibid., p. 7.

29   *Age*, 16 August 1928, p. 13.

30   *Sydney Morning Herald*, 18 August 1928, p. 17.

31   Ibid.

32   *The Australian Ski Year Book*, 1929, pp. 98–99.

33   *Sydney Morning Herald*, 17 August 1928, p. 11.

34   *Sydney Morning Herald*, 18 August 1928, p. 17.

35   Ibid.

36   *The Australian Ski Year Book*, 1929, p. 98.

37   *Referee*, 19 January 1939, p. 23.

38   *The Australian Ski Year Book*, 1928, p. 103.

39   *Sydney Mail*, 10 August 1927, p. 9.

40   Ibid.

41   *Sydney Mail*, 8 August 1928, p. 12.

42   *Labor Daily*, 8 August 1928, p. 4.

43   *Sydney Morning Herald*, 24 July 1928, pp. 14–15; *Evening News*, 23 July 1928, p. 12.

44   *Manaro Mercury, and Cooma and Bombala Advertiser*, 14 September 1928, p. 2.

45   Ibid.

46   *The Australian Ski Year Book*, 1929, pp. 98–99.

47   *Sun*, 14 August 1926, p. 7.

48   Ibid.

49   *Sydney Morning Herald*, 16 August 1928, p. 11.

50   *Manaro Mercury, and Cooma and Bombala Advertiser*, 14 September 1928, p. 2.

51   *Sydney Morning Herald*, 17 August 1928, p. 11.

52   Ibid.

53   Ibid.

54   *Sydney Morning Herald*, 28 January 1928, p. 18; *Sun*, 28 January 1928, p. 7.

55   *Sydney Morning Herald*, 17 August 1928, p. 11.

56   *Sydney Morning Herald*, 16 August 1928, p. 11.

57   *Goulburn Evening Penny Post*, 16 August 1928, p. 2.

58   *Sun*, 16 August 1928, p. 13.

59   *Evening News*, 16 August 1928, p. 11.

60   *Evening News*, 16 August 1928, p. 1.

61   *Evening News*, 16 August 1928, p. 11.

62   *Sun*, 16 August 1928, p. 24.

63   *Sunday Times*, 19 August 1928, p. 1.

64   Ibid.

65   *Herald*, 16 August 1928, p. 1; *Argus*, 17 August 1928, p. 11; *Age*,
      17 August 1928, p. 10; *Register*, 17 August 1928, p. 9; *Advertiser*,
      17 August 1928, p. 14.

66   *Daily Standard*, 16 August 1928, p. 9; *Telegraph*, 16 August 1928, p. 9;
      West Australian, 16 August 1928, p. 17; *Mercury*, 17 August 1928,
      p. 9; *Canberra Times*, 18 August 1928, p. 1; *Northern Territory Times*,
      17 August 1928, p. 5; *Mirror*, 18 August 1928, p. 2.

67   *Townsville Daily Bulletin*, 17 August 1928, p. 3; *Daily Advertiser*, 18
      August 1928, p. 1; *Border Watch*, 21 August 1928, p. 1.

68   *Auckland Star*, 16 August 1928, p. 7; *Evening Post*, 16 August 1928,
      pp. 11, 13.

69   *Brooklyn Daily Eagle*, 28 August 1928, p. 2.

Chapter Five   Search and Rescue

1   *Sydney Morning Herald*, 5 June 1909, p. 13.

2   Ibid.

3   *Sydney Morning Herald*, 5 May 1951, p. 9.

4   *Goulburn Evening Penny Post*, 27 January 1916, p. 4.

5   *Queensland Times*, 28 December 1915, p. 2.

6   *Sydney Morning Herald*, 10 May 1916, p. 9.

7   Ibid.

8   *Sydney Morning Herald*, 20 February 1918, p. 7.

9   *Northern Star*, 6 June 1918, p. 2.

10   Ella Airlie, 'Back to Kosciusko' (Sydney: W. H. Paling & Co., c. 1917).

11   *The Murrumbidgee Irrigator*, 23 April 1926, p. 5.

12   *Barrier Miner*, 27 September 1915, p. 4; *Manaro Mercury, and Cooma
      and Bombala Advertiser*, 23 April 1926, p. 2; *Sydney Morning Herald*, 26
      April 1926, p. 10.

13   *Daily Telegraph*, 22 August 1919, p. 6.

14   *Manaro Mercury, and Cooma and Bombala Advertiser*, 14 September
      1928, p. 2.

15   *Kyogle Examiner*, 18 June 1926, p. 5.

16   Ibid.

17   Ibid.

18   *The Australian Ski Year Book*, 1929, p. 100.

19   *Sydney Morning Herald*, 18 August 1928, p. 17.

20   *Sydney Morning Herald*, 16 August 1928, p. 11; 17 August 1928, p. 11.

21   *Sydney Mail*, 3 July 1929, p. 3.

22   *Sydney Mail*, 10 August 1927, p. 8.

23   *The Australian Ski Year Book*, 1931, pp. 53–56.

24 *Sydney Mail*, 10 August 1927, p. 8.

25 *Sydney Morning Herald*, 8 August 1925, p. 16.

26 *Sydney Morning Herald*, 16 August 1928, p. 11; 17 August 1928, p. 11.

27 *Argus*, 18 August 1928, p. 17.

28 *Sydney Morning Herald*, 18 August 1928, p. 18.

29 *Sydney Morning Herald*, 17 August 1928, p. 11.

30 *Sydney Morning Herald*, 18 August 1928, p. 17.

31 Ibid.; *The Australian Ski Year Book*, 1929, p. 99.

32 *Argus*, 28 August 1928, p. 17.

33 *Sun*, 17 August 1928, p. 11.

34 Ibid.

35 *Sydney Morning Herald*, 18 August 1928, p. 17.

36 *Sydney Morning Herald*, 19 January 1921, p. 11.

37 *Sun*, 10 August 1928, p. 11.

38 *Sunday Times*, 8 January 1928, p. 2.

39 Much of what follows comes from Greg Banfield, 'An interview with Rupert King', *The '14–'18 Journal*, 1991, pp. 64–71.

40 Banfield, 'An interview with Rupert King', p. 65.

41 *Sydney Morning Herald*, 20 December 1927, p. 11.

42 National Archives of Australia, B2455, SHIERS W. H. [8083162].

43 *National Geographic Magazine*, Vol. 39, No. 3, March, 1921, pp. 229–39.

44 *Manaro Mercury, and Cooma and Bombala Advertiser*, 17 August 1928, p. 3.

45 Banfield, 'An interview with Rupert King', p. 70.

46 *Sydney Morning Herald*, 18 August 1928, p. 17.

47 *Sun*, 17 August 1928, p. 11.

48 *Sydney Morning Herald*, 18 August 1928, p. 17.

49 Ibid.

50 Ibid.

51 Ibid.

52 Ibid.

53 Ibid.

54 *Sun*, 17 August 1928, p. 11.

55 Ibid.; *Telegraph*, 17 August 1928, p. 8.

56 *Sun*, 17 August 1928, p. 11.

57 *Sydney Mail*, 14 December 1932, p. 2.

58 Ibid.

59 *Bombala Times*, 12 January 1934, p. 2.

60 *Sydney Mail*, 14 December 1932, p. 2; *Bombala Times*, 12 January 1934, p. 2.

61 *Sydney Morning Herald*, 17 March 1906, p. 6.

62 *Sydney Mail*, 14 December 1932, p. 2.

63 *Sydney Mail*, 25 January 1932, p. 2.

64 *Daily Telegraph*, 6 August 1900, p. 9.

65 *Bombala Times and Monaro and Coast Districts General Advertiser*, 11 November 1904, p. 3; *Delegate Argus and Border Post*, 12 November 1904, p. 8.

66 *Aborigines Protection Act 1909* (NSW: *No. 25, 1909*).

67 *Bombala Times and Monaro and Coast Districts General Advertiser*, 20 December 1901, p. 3.

68 *An Act to prohibit the supply of Intoxicating Liquors to the Aboriginal Natives of New South Wales,* (NSW: *31 Vic. No. 16, 1867*).

69 *Goulburn Evening Penny Post*, 30 April 1928, p. 2; *Sydney Morning Herald*, 31 March 1928, p. 18; *Sydney Morning Herald*, 11 August 1930, p. 7.

70 *Sydney Morning Herald*, 11 August 1930, p. 7.

71 *Bombala Times*, 30 March 1928, p. 1.

72 *Goulburn Evening Penny Post*, 30 April 1928, p. 2.

73 *Bombala Times*, 30 March 1928, p. 1.

74 *Goulburn Evening Penny Post*, 1 May 1928, p. 3.

75 *Sunday Times*, 19 August 1928, p. 1.

76 *Sydney Morning Herald*, 17 August 1928, p. 11.

77 *Evening News*, 18 August 1928, p. 6.

78 *Sunday Times*, 19 August 1928, p. 1.

79 Ibid.

80 *Sydney Morning Herald*, 20 August 1928, p. 12.

81 *Sun*, 20 August 1928, p. 9.

82 *Sunday Times*, 19 August 1928, p. 1.

83 Ibid.

84 *Sydney Morning Herald*, 20 August 1928, p. 11.

85 *Sunday Times*, 19 August 1928, p. 1. The report only identifies 'Dr Teece', but it is clearly Lennox from wider context.

86 Ibid.

87 *Sun*, 20 August 1928, p. 9.

Chapter Six A Summit Too Far

1 *Daily News* (New York), 29 August 1928, p. 148.

2 *The National Geographic Magazine*, 30(6), December 1916, p. 476.

3 Ibid.

4 *Evening News*, 3 August 1925, p. 9.

5   *The Guardian*, 5 August 1925, p. 4.

6   For instance: *Princeton Daily Clarion*, 14 September 1925, p. 6; *Tampa Tribune*, 17 September 1925, p. 14; *Muncie Evening Press*, 21 September 1925, p. 6.

7   *Sydney Morning Herald*, 8 February 1902, p. 7.

8   *New York Times*, 19 February 1907, p. 10.

9   *Sydney Morning Herald*, 20 August 1928, p. 11.

10   Ibid.

11   Ibid.

12   Ibid.

13   *Sydney Morning Herald*, 22 August 1928, p. 15.

14   *Sun*, 20 August 1928, p. 9.

15   Ibid.

16   *The Australian Ski Year Book*, 1929, pp. 107–8.

17   *Canberra Times*, 30 July 1977, p. 11.

18   Ibid.

19   *Sydney Morning Herald*, 1 August 1927, p. 11.

20   *Sydney Mail*, 31 August 1927, pp. 12–13.

21   *The Australian Ski Year Book*, 1929, p. 107.

22   *Manaro Mercury, and Cooma and Bombala Advertiser*, 17 December 1920, p. 2.

23   *The Australian Ski Year Book*, 1929, p. 107.

24   Ibid, pp. 107–08.

25   Ibid, p. 108.

26   *Sydney Morning Herald*, 20 August 1928, p. 11.

27   Ibid.

28   Ibid.

29   *Sydney Morning Herald*, 21 August 1928, p. 11.

30   Ibid.

31   Ibid, p. 15.

32   *Sydney Morning Herald*, 20 August 1928, p. 12.

33   Ibid., p. 15.

34   Ibid., p. 13.

35   *Sun*, 20 August 1928, p. 2.

36   *Sun*, 21 August 1928, p. 15.

37   Ibid.

38   Ibid.

39   *Sydney Morning Herald*, 24 August 1928, p. 13.

40   Ibid.

41   *Canberra Times*, 13 June 1982, p. 7.

42 *Sunday Times*, 2 September 1928, p. 26.

43 *Sydney Morning Herald*, 21 August 1928, p. 11.

44 *Sydney Morning Herald*, 22 August 1928, p. 15; 23 August 1928, p. 11.

45 *Sydney Morning Herald*, 24 August 1928, p. 13.

46 *Sydney Morning Herald*, 27 August 1928, p. 11.

47 *The Land*, 24 August 1928, p. 1.

48 *Singleton Argus*, 23 August 1928, p. 3.

49 Ibid.

50 The *Singleton Argus* article was reprinted by the *North West Courier*, 30 August 1928, p. 6.

51 *Sydney Morning Herald*, 25 August 1928, p. 16.

52 *Manaro Mercury, and Cooma and Bombala Advertiser*, 31 August 1928, p. 2.

53 *Sunday Times*, 2 September 1928, p. 26.

54 *Freeman's Journal*, 13 September 1928, pp. 26–31.

55 *Manaro Mercury, and Cooma and Bombala Advertiser*, 14 September 1928, p. 2; *Sydney Morning Herald*, 10 September 1928, p. 13.

56 *Manaro Mercury, and Cooma and Bombala Advertiser*, 14 September 1928, p. 2.

57 Ibid.

58 Ibid.

59 *Bombala Times*, 12 January 1934, p. 2.

60 *Sydney Morning Herald*, 11 September 1928, p. 10.

61 *Manaro Mercury, and Cooma and Bombala Advertiser*, 14 September 1928, p. 2.

62 *Sun*, 12 September 1928, p. 15.

63 *Sydney Morning Herald*, 13 September 1928, p. 9.

64 *Sun*, 13 September 1928, p. 14.

65 Ibid.

66 Ibid.; *Sydney Morning Herald*, 14 September 1928, p. 17.

67 *Sun*, 13 September 1928, p. 14.

68 *Sydney Morning Herald*, 25 August 1928, p. 16.

69 *Smith's Weekly*, 25 August 1928, p. 1.

70 Ibid., p. 3.

71 Ibid.

72 Ibid.

73 Ibid.

74 *Armidale Chronicle*, 1 September 1928, p. 4.

75 Ibid.

76 Ibid.

77 *Glen Cove Echo*, 31 August 1928, p. 1.
78 Ibid.
79 *Brooklyn Daily Eagle*, 28 August 1928, p. 2.
80 *New York Times*, 12 September 1928, p. 9.
81 *Evening News*, 20 September 1928, p. 1.
82 *Evening Post* (New Zealand), 20 September 1928, p. 11; *Telegraph*, 20 September 1928, p. 9.
83 *Canberra Times*, 13 June 1982, p. 7, quoting the *Glen Cove Echo*.
84 *Age*, 16 October 1928, p. 10.
85 Ibid.
86 *Sydney Morning Herald*, 13 October 1928, p. 16.
87 *Sydney Morning Herald*, 15 October 1928, p. 11.
88 *Canberra Times*, 13 June 1982, p. 7.
89 *Sun*, 3 October 1928, p. 5.
90 Ibid., p. 12.
91 Ibid., p. 2.
92 *Canberra Times*, 13 June 1982, p. 7.

Chapter Seven   A Vital Clue
1 *Sun*, 17 August 1928, p. 11.
2 *Daily Telegraph*, 23 August 1927, p. 10.
3 *Daily Telegraph*, 8 June 1927, pp. 1, 14–15.
4 *Daily Telegraph*, 25 June 1927, p. 18.
5 *Daily Telegraph*, 23 July 1927, p. 1.
6 *Daily Telegraph*, 20 July 1927, pp. 16–17.
7 *Sydney Mail*, 3 August 1927, p. 5.
8 *Daily Telegraph*, 23 July 1927, pp. 22–23.
9 *Sydney Morning Herald*, 19 September 1928, p. 15.
10 *Sydney Morning Herald*, 16 August 1911, p. 20.
11 *Fitzroy City Press*, 25 November 1897, p. 3.
12 *Kalgoorlie Miner*, 18 April 1898, p. 6.
13 *Sydney Morning Herald*, 19 September 1928, p. 15.
14 Ibid.
15 Ibid.
16 *Sydney Morning Herald*, 1 December 1928, p. 18.
17 *Sydney Morning Herald*, 24 April 1928, p. 6.
18 *Sydney Morning Herald*, 1 December 1928, p. 18.
19 *Sydney Morning Herald*, 17 December 1928, p. 8.
20 Ibid.
21 Ibid.; see also *Sydney Morning Herald*, 12 January 1929, p. 22.

22 *Evening News*, 12 January 1929, p. 3.

23 Ibid.

24 *Age*, 19 December 1928, p. 18.

25 *Evening News*, 11 January 1929, p. 1.

26 *Wagga Wagga Advertiser*, 17 January 1877, p. 2.

27 *Manaro Mercury, and Cooma and Bombala Advertiser*, 4 February 1929, p. 2.

28 *Sydney Morning Herald*, 4 April 1929, p. 15.

29 Ibid.

30 *Sydney Morning Herald*, 1 December 1928, p. 18.

31 *Sydney Morning Herald*, 22 December 1928, p. 10.

32 *Daily Telegraph*, 19 December 1906, p. 8.

33 *Sunday Times*, 7 June 1925, p. 7.

34 *Sydney Mail*, 18 December 1929, p. 11.

35 *Manaro Mercury, and Cooma and Bombala Advertiser*, 4 January 1929, p. 3.

36 Ibid.

37 *Illawarra Mercury*, 29 March 1929, p. 1.

38 *Sydney Morning Herald*, 4 April 1929, p. 15.

39 *Sydney Morning Herald*, 18 May 1928, p. 17.

40 Ibid.

41 *Sydney Morning Herald*, 20 May 1929, p. 12.

42 Ibid.

43 Ibid.

44 *Sydney Morning Herald*, 18 May 1929, p. 17; 20 May 1929, p. 12.

45 The above reports say Laurie's parents, but *Sun*, 18 May 1929, p. 3, says it was from Chrissie.

46 *Sydney Morning Herald*, 20 May 1929, p. 12.

47 *Manaro Mercury, and Cooma and Bombala Advertiser*, 27 May 1929, p. 2.

48 *Sydney Morning Herald*, 9 January 1929, p. 14.

49 *News*, 21 August 1928, p. 1.

50 *Sun*, 18 May 1929, p. 3.

51 *Sun*, 13 February 1929, p. 21.

52 *Evening News*, 16 March 1929, p. 8; *Sun*, 18 March 1929, p. 15.

53 *Sydney Morning Herald*, 4 May 1929, p. 11; *Evening News*, 6 May 1929, p. 10; *Sun*, 16 June 1929, p. 32; *Sydney Morning Herald*, 17 June 1929, p. 4; *Evening News*, 15 July 1929, p. 12; *Sydney Morning Herald*, 15 July 1929, p. 4.

54 *Sydney Morning Herald*, 17 April 1929, pp. 17–18.

55   *Advertiser*, 22 May 1929, p. 15.

56   *Sun*, 15 December 1929, p. 38.

57   *Sydney Morning Herald*, 1 January 1930, p. 7. A different account had the pair as cattlemen: *Manaro Mercury, and Cooma and Bombala Advertiser*, 3 January 1930, p. 2.

58   *Sydney Morning Herald*, 1 January 1930, p. 7.

59   *Manaro Mercury, and Cooma and Bombala Advertiser*, 3 January 1930, p. 2.

60   *Sydney Morning Herald*, 2 January 1930, p. 10.

61   *Sydney Morning Herald*, 3 January 1930, p. 13.

62   *Evening News*, 2 January 1930, p. 6.

63   *Sun*, 2 January 1930, p. 4.

64   *Lithgow Mercury*, 10 January 1930, p. 4.

65   *Sydney Morning Herald*, 29 June 1937, p. 5; *Construction and Real Estate Journal*, 30 June 1937, p. 16.

66   *The Australian Ski Year Book*, 1929, pp. 156–8, 170–1.

67   Ibid., pp. 97–107.

68   *The Australian Ski Year Book*, 1930, p. 71.

69   *Evening News* (Rockhampton), 24 July 1933, p. 2.

70   *Sydney Mail*, 31 August 1938, p. 15.

71   *Smith's Weekly*, 14 June 1941, p. 17.

72   Ibid.

73   *Sun*, 28 December 1941, p. 5.

74   *Sydney Morning Herald*, 11 August 1951, p. 7.

75   *Canberra Times*, 12 June 1982, p. 16; *Canberra Times*, 13 June 1982, p. 7.

76   David Scott, 'Death of the Summit: The Seaman and Hayes tragedy of 1928 and the construction of the Seaman Memorial Hut', Paper for the Kosciuszko Huts Association, August 2013: https://khuts.org/images/stories/history/DeathOnTheSummit_dscott_13aug2013.pdf.

77   For other regional histories encompassing related subjects see the work of Matthew Higgins: *Bold Horizon: High-Country Place, People and Story* (Dural: Rosenburg, 2018); *Rugged Beyond Imagination: Stories from an Australian Mountain Region* (Canberra: National Museum of Australia Press, 2009); and of Peter Southwell-Keely: *Highway to Heaven: a History of Perisher and the Ski Resorts along the Kosciuszko Road* (Gordon: Perisher Historical Society, 2013); *Out on the tops: the centenary of the Kosciuszko Alpine Club* (Chatswood: Kosciuszko Alpine Club, 2009).

Chapter Eight   Myths and Realities

1   *Sydney Morning Herald*, 17 January 1902, p. 4; *Wellington Times*, 21 May 1925, p. 6.

2   *Freeman's Journal*, 27 February 1913, p. 25.

3   *Sydney Morning Herald*, 23 August 1930, p. 11; *Australasian*, 28 June 1930, p. 57.

4   *Sydney Morning Herald*, 10 July 1930, p. 10; *Sydney Mail*, 16 July 1930, p. 9; *Referee*, 16 July 1930, p. 26.

5   *Newcastle Sun*, 15 August 1934, p. 7.

6   *Newcastle Morning Herald and Miner's Advocate*, 2 October 1935, p. 6.

7   *Tumut and Adelong Times*, 24 March 1936, p. 6; *Advertiser* (Adelaide), 2 April 1936, p. 7.

8   *Sydney Morning Herald*, 16 January 1939, p. 15; *Referee*, 19 January 1939, p. 23.

9   *Bombala Times*, 12 January 1934, p. 2.

10 Ibid.

11 *Southern Districts Advocate*, 16 January 1933, p. 1.

12 *Sydney Morning Herald*, 24 January 1933, p. 10.

13 *News* (Adelaide), 5 May 1932, p. 12.

14 *Sun*, 19 January 1933, p. 28.

15 Ibid.

16 *Sydney Mail*, 12 April 1933, p. 40.

17 *Truth*, 7 May 1933, p. 22.

18 *Sun*, 9 August 1933, p. 17.

19 *Sun*, 15 October 1933, p. 29.

20 *Sun*, 28 December 1933, p. 12.

21 *Sun*, 7 January 1934, p. 23.

22 *Sydney Morning Herald*, 4 August 1934, p. 19.

23 *Truth*, 22 July 1934, p. 23; *Sun*, 23 September 1934, p. 29.

24 *Sun*, 18 February 1940, p. 4; *Sydney Morning Herald*, 19 February 1940, p. 13.

25 *Sydney Morning Herald*, 19 February 1940, p. 13.

26 *Sun*, 18 February 1940, p. 4.

27 *Sydney Morning Herald*, 27 January 1940, p. 13. This plaque is now in the Polish embassy.

28 Thomas Campbell, *The Pleasures of Hope with Other Poems*, New Edition (London: Longman, Hurst, Rees, Orme, and Brown, 1822) p. 38.

29 *Sydney Morning Herald*, 27 November 1939, p. 10.

30 For example: *Herald*, 7 October 1942, p. 1; *West Australian*, 31 October 1942, p. 5; *Canberra Times*, 12 November 1942, p. 3.

31   *Manning River Times and Advocate*, 4 October 1941, p. 4; *Yass Tribune-Courier*, 17 July 1944, p. 4.

32   *Daily Telegraph*, 29 August 1941, p. 6.

33   *Telegraph* (Brisbane), 31 May 1940, p. 2; *West Australian*, 3 June 1940, p. 6.

34   *Australian Worker*, 15 September 1943, p. 6.

35   Ibid.

36   *Kosciusko State Park Act* (NSW: *George VI, No. 14, 1944*).

37   *Sun*, 16 March 1939, p. 10; *Macleay Argus*, 9 May 1939, p. 5; *Sun*, 18 June 1939, p. 5; *Lithgow Mercury*, 27 September 1939, p. 5;

38   *Sydney Morning Herald*, 1 November 1944, p. 7.

39   *Sydney Morning Herald*, 12 February 1925, p. 13.

40   *National Parks and Wildlife Act* (NSW: *Elizabeth II, No. 35, 1967*).

41   Paul Edmund de Strzelecki, *Physical Description of New South Wales and Van Diemen's Land* (London: Longman, Brown Green and Longmans, 1845), p. 62.

42   Elyne Mitchell, *Australia's Alps* (Sydney: Angus and Robertson, 1946), p. 67.

Acknowledgements

Each book is a journey and I'm grateful to those who help me make it. Thanks to the team at Hardie Grant Books for their continued support. Pam Brewster deserves special mention as *Kosciuszko*'s foremost mountain guide, discerning the book's potential and helping me see the journey through drafting and crafting. Thanks also to Marg Bowman and Tricia Dearborn for helping me pull the text into shape, and to Dale Campisi for his guidance and positivity during the final push. And thanks to the Hardie Grant Books publicity team for their hard work, and to Nada Backovic for another compelling cover.

As always, I want to acknowledge those major institutions that hold our evidence in trust, most especially the National Library of Australia, the state libraries of New South Wales and Victoria, and the National Archives of Australia. Thanks also to Sydney's Northern Beaches Library and the Glen Cove Public Library of New York. I also want to commend the Thredbo Ski Museum, where Laurie's camera remains an important sign of the global significance of local history.

I'm thankful to those companions on the way who make historical research and writing a little less lonely an occupation, whether family or friends, near or afar. I'd like to single out Pru Francis and Eloise Armstrong at the Archdiocese of Hobart Archives as especially supportive colleagues, and Michael Tate and Anthony Ray for more than a few long blacks. Susan Brodie, Sophia Brodie and Kristyn Harman all deserve special mention for accompanying me to the top of Kosciuszko. Kristyn earns further notice for enduring the many months of research and writing that distracted me on either side of the summit.

Finally, I'd like to close by honouring the memories of Evan and Laurie and all those affected by the tragedy of 1928. I hope I have done their stories justice.